# Genesis-aligned Biochemistry

## The End of Atheism

### (Revised Addition)

# Genesis-aligned Biochemistry:
# The End of Atheism

Alan J Parsonage



ISBN: 978-1-9195048-3-4

Published by SynergyWorking. This is available to order from most book shops in the United Kingdom and is globally available via internet book retailers andwww.amazon.com, Bookvault and IngramSpark.

Copy-edited by Graham Hughes

Cover design by inspiration of Suzanne Parsonage and produced by Stefan Proudfoot of Spiffing Publishing.

Bible verses and search and study tools are from www.biblegateway.com and www.blueletterbible.org.



The main ideas of this book are from myself, but they were certainly aided by my recognition of one modern-day prophet, Robin D Bullock, and one modern-day revelator, Kat Kerr; their insights led to mine. If any of their insights cannot be confirmed with scripture, I will reference them with a superscript endnote: endnote[i] or [KK] and endnote[ii] or [RDB]. There are more specific references to their insights in my book *God's Kosmos-Ruah Metaphor.*

Significant Bible verses will be in the body of the text and ancillary ones will be in footnotes. Long biblical passages or previously footnoted Bible verses will just be referenced and not quoted.

***'I believe God made me for a purpose, but he also made me fast and when I run I feel his pleasure'*** **[1, 2]**

**(For me to 'run' was pattern recognition)**

**For Suz**

**For Becky, Nick and Nelly**

---

[1] A quote from the character playing the real-life person Eric Liddell in the film *Chariots of Fire* directed by Hugh Hudson.

[2] *'The parable of the talents'*: Matthew 25:14-30 describes God's pleasure when Man uses his given abilities to the full.

# Table of Contents

# Part 1

## Genesis-day-aligned Biochemistry

God used the same framework of design when inspiring the book of Genesis and in creating the biochemistry of all life – a demonstrable proof of a shared designer.

All languages were formed with the Bible in mind.

The Bible often uses metaphors and symbolism to convey spiritual truths – Jesus Christ declared: *'I am the true vine'*.

The flow and chronology of Genesis, point by point, parallels biochemistry – compound by compound, reaction by reaction, and process by process.

Elements and biochemical compound names will by their etymology, when viewed in a biblical context, reveal the role of an 'entity' they represent within a certain aspect of the salvation process. That entity will be one of, or a combination of: God the Father; the Lord Jesus Christ; the Holy Spirit; the Bible; Man's original spirit; Man's regenerated spirit; or any image or likeness given by God.

# Chapter 1

# The Kosmos-Ruah Metaphor

## *What is the Kosmos-Ruah Metaphor?*

This book is an adjunct to my book *God's Kosmos-Ruah Metaphor* (the 'KRM book'); and the KRM book describes the Kosmos-Ruah Metaphor ('the KRM') in much more detail than here. In order to understand the links between Genesis and the biochemistry of all life on Earth, a certain amount of understanding of the KRM is required, so this book will briefly explain it, and contains several quotes and four close-copy entire chapters from the KRM book.

> '*Kosmos*' (Greek) – *what adorns the Earth, or our world*
>
> '*Ruah*' (Hebrew) – *the 'spirit man' of Man*

**The KRM is a prolonged and very precise metaphor with multiple scriptural and logical 'typing' (metaphorising) that runs throughout Genesis 1 and 2. It tells the story of mankind's redemption by the Abrahamic Covenant and after that through the Lord Jesus Christ – every single verse and every single word of Genesis 1 and 2 are not just about God making the world but also prophesy these redemptive plans in great detail. I will show how this metaphor dominates the design of our natural world.**

Once the KRM is understood, an explosion of new insight is revealed: every fossil layer is explained; the K-T extinction level becomes predictable; the existence and size of prehistoric dinosaurs, sea reptiles and trees are shown to have a biblical 'reason'; the Precambrian and the Cambrian are explained; a plausible explanation for the apparent great ages produced by radiometric dating becomes evident; and even Darwin's Galapagos finches can be viewed as testifying, very specifically, to Christ's gifts to saved men.

The Genesis text sets up several 'compare and contrasts' within this metaphor and shows incredible skill to do so. This makes for a 'doubly constructive' nature that reveals extremely precise comparisons: before- and after-salvation conditions; images to likenesses; Adam compared to Jesus Christ; pre-fallen Adam compared to saved Man; the Abrahamic Covenant compared to the better covenant of Christ; and all of this enclosed within an account of how God made the world. It becomes apparent that God specifically scheduled the world-making process in such a way as to tell these other narratives, but did He make other aspects of His creation reveal the same story?

The parallels are introduced by Paul:

> ***2 Corinthians 4:6** (KJV): **For God, who commanded the light to shine out of darkness, hath shined in our hearts, to give the light of the knowledge of the glory of God in the face of Jesus Christ.***

Paul the Apostle knew of a parallel between God commanding light to shine on a world described in Genesis 1:2 as: '*without form and void*' ('*tohuw*' and '*bohuw*') and with its waters 'covered' by darkness; comparing it to God shining light on Man's darkened heart (his spirit and soul) – thereby introducing a parallel or metaphor of the whole world with Man's spirit and soul. Also, what God makes 'available' – the light – then represents what God Himself gives – His actions. The light '*shined*' and gave knowledge – God gives knowledge with His light – something from God is given to Man. The KRM always follows the same format: something for the whole salvation process and then something for the individual Man's spirit.

It is with just one Hebrew word by which the Genesis text is unlocked (as never before) – that word is *'qara'* (*'qara' – proclaim, call, call oneself*). *'Qara'* is used nine times in Genesis 1 and 2. In Genesis 2:23[3] we see it used to name (*'call'*) a woman a *'woman'* because she *'came out of man'*. I put emphasis on *'called, because she came out of'*, and something was now in someone else and not on the name *'woman'*. The gist being, if it comes out of you and is 'of you' but now 'of' or 'in someone else' – then you 'proclaim' it.

The first five other uses of *'qara'* occur in Days 1 to 3 and it is not used in Days 4 to 6. I suggest, within the KRM that in each case where *'qara'* is used, it is used with terms that represent the Trinity God's 'nature' or 'substance' coming from God and now being in Man, just like *'light'* in 2 Corinthians 4:6. Although the words have meaning in the literal context it is within the 'prolonged metaphor' that this second meaning has more significance. God *'gave names'* to, or *'called'* (*'qara'*), in Genesis 1: ***'darkness'*** – *'night'*, ***'light'*** – *'day'*, ***'firmament'*** – *'heaven'*, ***'waters'*** – *'seas'* and ***'dry land'*** – *'earth'* (*'heaven'* can mean an abode of God). All except *'darkness'* could be interpreted as representing something being given to Man (like God giving *'light'* in the Corinthians passage). Even *'darkness'* (evil) could be considered to be 'from God' because it is revealed by His light that *'divided the light from the darkness'*; hence its detection and 'existence' is also from God.

To understand how much importance God puts on the KRM, I am going to give one example of it and show how many times God repeats this same story – it is repeated throughout all of Genesis 1 and 2. It is this same story that appears in the biochemistry of all life (in the correct chronological order) and in ways that will astonish you.

---

[3] Genesis 2:22-23 (KJV): ***And the rib, which the Lord God had taken from Man, made he a woman, and brought her unto the Man. And Adam said, this is now bone of my bones, and flesh of my flesh: she shall be called*** *('qara')* ***woman because she was taken out of Man.***

**Day 3 of Genesis and the relevant metaphor key**

**From scripture:** *'Grass and herbs'* are used in the Bible as a metaphor for Man. Grass is often likened to *'flesh'* (Man and animals are included in *'flesh'* (1 Peter 1:24[4])); and the repetitive uses of *'herbs'* and *'grass'* (again) both symbolise Man (2 Kings 19:26[5]). *'Seed'* is often used to refer to the redeemed, the Word of God (the Bible) or, perhaps, the indwelling 'substance' of God given to Man (Matthew 13, Mark 4:26[6], Galatians 3:16[7]). The Holy Spirit is described as a necessity to be fruitful ('fruitful': to show Christ- or Holy Spirit-like behaviour; give knowledge of Christ; and give spiritual help (John 15:4[8] and Romans 8)).

**From logic:** God can only use the vocabulary of that day to establish His metaphor (this may be plants, lights, fish, birds or animals). Please note the 'two sets of three' pattern in Genesis 1:11-12.

> ***Genesis 1:11-12** (KJV): **11. And God said, Let the earth bring forth** ('dasa') **grass** (1), **the herb yielding seed** (2), **and the fruit tree** (3) **yielding fruit after his <u>kind</u>, whose seed is in itself, upon the earth: and it was so. 12. And the earth brought forth** ('yasa') **grass** (1), **and herb yielding seed** (2) **after his <u>kind</u>, and the tree yielding fruit** (3), **whose seed was in itself, after his <u>kind</u>: and God saw that it was good.***

---

[4] 1 Peter 1:24 (KJV): ***For all flesh is as grass, and all the glory of man as the flower of grass. The grass withereth, and the flower thereof falleth away:***

[5] 2 Kings 19:26 (KJV): ***Therefore their inhabitants were of small power, they were dismayed and confounded; they were as the grass of the field, and as the green herb, as the grass on the house tops, and as corn blasted before it be grown up.***

[6] Mark 4:26 (KJV): ***And he said, So is the kingdom of God, as if a man should cast seed into the ground.***

[7] Galatians 3:16 (KJV): ***Now to Abraham and his seed were the promises made. He saith not, And to seeds, as of many; but as of one, And to thy seed, which is Christ.***

[8] John 15:4 (KJV): ***Abide in me, and I in you. As the branch cannot bear fruit of itself, except it abide in the vine; no more can ye, except ye abide in me.***

As *'herbs'* are 'typing' Man and *'seed'* is the redeemed, then *'the herb yielding seed'* of Genesis 1:11-12 is a logical descriptive metaphor of Jesus as a man yielding the redeemed (*'seed'*) by His work. This term can also apply to every saved man, who yields his own *'seed'*.

Given that trees are supernatural educators (Genesis 2:9[9]), the Holy Spirit is described as fruitful[10] and the tree may be looked upon as the biggest plant available (I will later show how this use of size to represent God is shown with the metaphorising of *'whales'* in the KRM) – then *'the fruit tree yielding fruit'* is a logical descriptive metaphor for the Holy Spirit and His works and teachings.

**Please note the differences between Verse 11 and Verse 12 (these differences describe before- to after-salvation changes):** they each use different words translated as 'brought forth' (*'yasa'* and *'dasa'*); the *'herb yielding seed'* has a definite article 'the' in Verse 11 but not 12; the *'herb yielding seed'* is not a *'kind'* in Verse 11 but is in Verse 12; *'the fruit tree'* becomes *'the tree'*, changes from having a seed *'is in itself'* to a seed that *'was in itself'* and is no longer *'upon the earth'*; again from Verse 11 to 12 respectively. Please also note that there is no 'God made' or 'God created' statement. All of these subtle differences or omissions have very specific meanings within the KRM.

**A three-day reverence** (darkness becomes light after three days)

Please note that seed-bearing life (representing the saved) begins three days after the whole world was in darkness – this is an example of a 'three-day reverence' – Christ's three days in the grave acclaimed.

---

[9] Genesis 2:9 (KJV): ***And out of the ground made the Lord God to grow every tree that is pleasant to the sight, and good for food; the tree of life also in the midst of the garden, and the tree of knowledge of good and evil.***

[10] Galatians 5:22-23 (KJV): ***But the fruit of the Spirit is love, joy, peace, longsuffering, gentleness, goodness, faith, meekness, temperance: against such there is no law.***

**The mechanics of the salvation of Man**

This means that **just** from Genesis 1:11-12 alone, I can deduce (or we can say the text prophesies) that God sends forth Man, Jesus Christ and the Holy Spirit carrying the Word of God (God sends forth four things). There will be a Holy Spirit interaction with Man. The to-be-saved Man will accept the offer of the Holy Spirit and will be indwelt with both the Lord Jesus Christ and the Holy Spirit. The saved Man will gain various other spiritual gifts too, and in the future, he will spiritually grow, become more Christ-like and help in evangelising others – a nice summary of the New Testament – all just from these two verses. **But here is the amazing thing: God repeats this over and over again; this story is the 'big deal'.**

Genesis 1:14-18, Genesis 1:20-23, Genesis 1:24-25, Genesis 1:26-28, Genesis 1:29-30, Genesis 1:29 to Genesis 2:1, Genesis 2:5-7, Genesis 2:5-18, Genesis 2:19-20 and Genesis 2:21-25 all tell exactly the same story. Genesis 1:29, Genesis 1:30 and Genesis 2:19 could also be interpreted to tell this same story as individual verses. On top of that, the 31 verses of Genesis 1:2 to Genesis 2:2 again tell exactly the same overall story, as does Genesis 2:5-25. In addition, Genesis 2:2-3 provides a comparison between the Abrahamic Covenant and the better covenant of Christ. Genesis 2:4 provides a comparison of Adam (and how all men were meant to be) with saved men. In Genesis 1, Days 1 to 3 describe the roles of the individual members of the Trinity in both the Abrahamic and better covenant with Christ, and Genesis 1:1-2 describes Adam's fall and spiritual condition. Genesis 1:2 also describes the Holy Spirit initiating redemption, and the whole of Genesis 1 can be interpreted as describing God's plan changing from the Abrahamic Covenant to the better covenant of Christ. There is also a cross-match between Genesis 1 and Genesis 2 in that Genesis 2:5-6 matches with Genesis Days 1 to 3, Genesis 2:8-9 matches with Genesis Day 4, Genesis 2:9-18 matches with Genesis Day 5, and Genesis 2:19-

25 matches with Genesis Day 6. All of these are, quite amazingly, in the correct chronological order, and all of the supposed contradictions in Genesis disappear once this is understood. (I will often refer to chronology as my 'proof'; it is all very well having a precise metaphor, but things happening in the correct order I find the most compelling.)

With this knowledge, all of the supposed 'flowery and poetic' terms used in the KJV of Genesis are shown to be very skilfully crafted and all 'slanted' to have a second meaning. Thus, *'lights'*, *'greater light'*, *'lesser light'*, *'a moving creature that has life'*, *'a living creature that moves'*, *'fowl'*, *'winged fowl'*, *'fowl of the air'*, *'the living creature'*, *'creeping thing'*, *'thing that creepeth on the earth'* and *'fish of the sea'* all have very, very specific and scripturally logical second meanings. I will gradually introduce the metaphoric meanings of these terms throughout this book.

It is not just this repeating story of salvation that I will demonstrate to be present in biochemistry but also the actual Genesis text itself; this is the second 'big deal'. Astoundingly, I will show how every pertinent point in the Genesis text will have a biochemical compound or interaction to represent it, and with the exact same chronology and flow as the Genesis text.

**Image to likeness**

In Genesis 1:26 God declares He will make Man *'in our image and after our likeness'* – in that particular order. In consistency, Genesis 1:27 describes images and Genesis 1:28 describes likenesses – the order is maintained (broadly, 'images' are substance and 'likenesses' are skill sets, behaviours and identities, but I will define images and likenesses in much more detail in the next chapter). As God is making Man's born-again spirit in the KRM, He uses the same structuring rule and always describes it in an 'image-then-likeness' order. Returning to Genesis 1:11-12, the Man (*'grass'*) and Jesus Christ (*'herb yielding seed'*) are both images of God and the **working** Holy Spirit (*'fruit tree'*) is a likeness and the order is maintained. Image-to-likeness formatting occurs throughout Genesis 1 and 2, always in the same order within sentences. This is one of seven structural rules in total within the KRM and in a way they confirm its validity. The last of these has staggering complexity, and I will reveal one aspect of it with something I have named a 'birthing sequence' – 'one true pattern' – later in this book.

The structural rules I will focus on are: 'chronology', 'before and after salvation', 'image to likeness' and 'three-day reverence'. These will all feature prominently in the biochemistry of life.

Please remember that God sends forth four entities: Man, Jesus Christ, the Holy Spirit and the Word of God (the Bible). They all have specific roles: Jesus accomplishes salvation by His gift and the Holy Spirit works with the Bible and is the 'promoter', 'project manager' and 'workforce' for the whole process. Also, within the KRM and the New Testament itself, you could say there is a kind of 'complementary coupling' between these entities – the Bible with the Holy Spirit, and Jesus with Man. Please also remember that many biochemical compounds exist in both left-handed (L) and right-handed (D) forms (mirror images of each other like our hands), and this property is called chirality.

**Not actually research but a little research like**

I started to write this book as an observational study with a research question of: 'do parallels with the hidden metaphor in Genesis exist in the biochemistry and physiology of all life?' My main 'rule' was simple: are biblical truths and chronologies in some way present in our micro world. (I will reveal several other 'rules' as we go along.)

On realising that these parallels were everywhere I looked, the research question quickly changed to: 'is there so much evidence of a shared design that we now have a (previously elusive) demonstrable proof of God?'

# Chapter 2

# God's Own DNA

*DNA – the building blocks of life*
*DNA – the building blocks of eternal life*

Some say that as much as 30% of the Bible is in the form of symbolism and metaphors, and in Genesis 1 and 2, Paul the Apostle helped to reveal a key that showed how plants, lights, fish, birds and animals metaphorised God giving something to Man at salvation. This was confirmed by a series of seven structural rules. Does DNA, the most famous biochemical molecule of them all, do the same and is this confirmed by 'formulaic' rules? **First comes the theory, then the proof and then the pattern.**

## The Theory: Metaphoric Nucleotidal Waves

***... all things were created by him (*Christ*) and for him'*** (Colossians 1:16)

With this verse in mind, I am going to look for specific instances of Christ-inspired design drivers. I believe that the Holy Spirit stands to the left of God the Father,[KK] and the Bible says Jesus is on His right (Mark 16:19[11]). The Lord Jesus Christ accomplished or 'made' salvation for Man, and the Holy Spirit applies or 'works' on salvation for Man – He 'gets the job done'. All life forms are made of left-handed or 'L'-amino acids (except glycine) and right-handed or 'D'-sugars, including the sugar 'ribose' present in DNA, RNA and adenosine triphosphate (ATP). This is

---

[11] Mark 16:19 (KJV): ***So then after the Lord had spoken unto them, he was received up into heaven, and sat on the right hand of God.***

termed 'homochirality', and it is a naturalistic mystery as to why this is so, as the other-handed (or mirror-image) options exist. As amino acids make up all life and as sugars are the energy supply for almost all life too, are these compounds then reflecting the roles of Jesus (the maker) and the Holy Spirit (the actioner) in making all life and Man's born-again spirit? In both cases the chirality of the chemical compounds 'points back to' or 'reflects' or 'proceeds from' the 'middle' or God the Father. This homochirality could be being used just like '*qara*' in Genesis 1 and 2 in that L- and D-chiral compounds can be viewed as representing 'given by' or 'of' Jesus, and the 'working' Holy Spirit respectively.

With this consideration, similarities to Christ's work in Man's salvation seem to exist in the structure, actions and even chemical formula of DNA, RNA and ATP. I will show that the two-stranded 'living' DNA may be representing the dynamic relationship between God and saved Man – one strand each. Not only that, but their molecular structures and actions also seem to follow the same confirmatory, 'before and after', 'chronological', 'three-day reverence' and 'image-then-likeness' pattern found in Genesis 1 and 2.

**I suggest that the 'reason' for homochirality, in all of creation, is to testify to the Lord Jesus Christ and the work of the Holy Spirit – a 'themed' creation telling a second narrative through biochemistry.**

**Etymology** (God is in control)

Etymology is the study of the origin and evolution of words in terms of their meaning and form. Throughout this book I will demonstrate that the root-word meaning of many compounds is identical to: established biblical metaphors; specific roles of Man or God; or specific traits of Man or God. Please note the etymology of common chemical suffixes:

***-ine***: 'of' or 'pertaining to', 'of the **nature** of' or 'made of'
***-il***: 'of or pertaining to', '**derived** from' a base compound
***-ide***: 'a simple derived unit', '**a structural unit**' of a bigger compound
***-ate:*** 'having the quality of' or '**having been acted upon**'.

**A nucleotide**

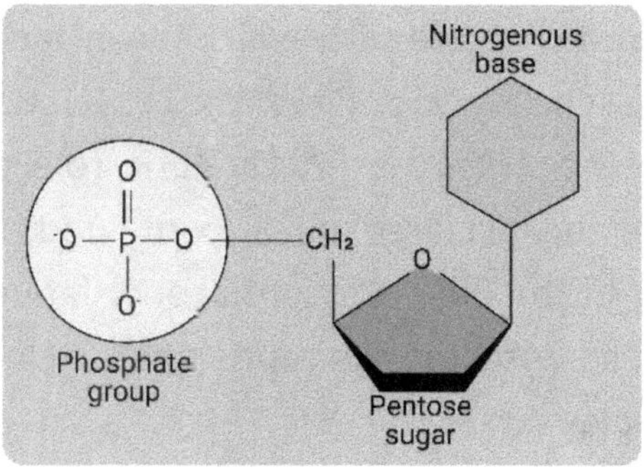


**Figure 2.1: A nucleotide; Jaya Bharathi A/Shutterstock.com**

DNA consists of repeating nucleotides that always consist of three parts (a 'set of three'): a phosphate group, ribose (a pentose sugar) and one of four nitrogenous bases – adenine (A), thymine (T), cytosine (C) and guanine (G). In RNA thymine is replaced with uracil (U). **The nucleotide is the 'unit' of this metaphor – each nucleotide represents a little 'package of Holy Spirit activity'. Please note that the original root-word meaning of *phosphorous* ('P' at the centre of the phosphate group) is 'light bringer' or 'light bearer'.**

The KRM and the New Testament describe God the Father purposing salvation ('God said') and reveal five participants in this process: Jesus Christ, the Holy Spirit, the Bible, Man's whole original spirit and the regenerated aspect of born-again Man's spirit. In Genesis Day 3, four of these are (metaphorically) represented as: Man ('*grass*'), Jesus Christ ('*the herb yielding seed*'), the Holy Spirit ('*the fruit tree*') and the Word of God or Bible ('*seed*'). Is there a similar and logical representation in nucleotide nitrogenous bases based on their etymological meanings? (The root meanings used from here on are from various etymological websites, but mainly Oxford Languages (via Google) and www.etymonline.com, and sometimes through Google AI and ChatGPT.) Please note the suffixes *-ine* and *-il* can mean 'pertaining to' and this, in a way, can also mean 'metaphorising'.

*Guanine* has etymology associated with 'bat dung' and 'seagull dung' which is a very powerful fertiliser – guanine is a fertile product that comes out of a creature of the air. This is a concise summary of the Holy Spirit's essential and defined role in salvation – so guanine can represent the Holy Spirit by a 'metaphoric role play'. *Cytosine* has etymology of an 'encapsulated' or a 'contained' thing and this is a concise summary of the Bible's essential and defined role in salvation – 'God in a book'; so, cytosine can represent the Holy Spirit by another 'metaphoric role play'.

*Adenine* has etymology associated with a 'gland' (or an 'acorn') that secretes (or gives) and as Christ is described as a 'seed' in Man and gives His life for, and a part of His spirit to Man, then this etymology is a concise summary of Christ's essential and defined role in salvation – so adenine can represent Christ by yet another 'metaphoric role play'. *Thymine* has etymology associated with the thymus gland which means 'spirit' or 'soul' – this is a concise summary of Man's essential and defined role in salvation – Man provides the spirit and soul (the 'most coveted real estate in the universe'[RDB]); so, thymine can also represent Man by a 'metaphoric role play'.

*Uracil* has etymology associated with a 'product of waste materials' and maybe specifically 'waste water' – this is a concise summary of what Man's born-again spirit is remade from (the '*tohu* and *bohu*' spirit, and '*darkness*' waters in Genesis 1:2) and so uracil can also represent the regenerated spirit of Man by a 'metaphoric role play'.

The five bases occur as 'complementary pairs' (in DNA and RNA) and with the same two coupling together: A with T (or A with U in RNA), and C with G. The **exact same coupling** occurs in the KRM (confirmed in the New Testament) as it is scripturally logical that the Holy Spirit is linked with the Word of God (as He inspired it and interprets it for us); and Jesus Christ is obviously linked with both Man and the regenerated aspect of Man, as He became one, and 'all' saved men are in the 'body of Christ'. A and G are called 'purines', and this has a meaning of 'pure ones', which again seems apt in referencing God. C and T are called 'pyrimidines', and this has a literal meaning of 'flame containing vessel', which also seems befitting if the flame is looked on as 'light'. Thus, the five participants in the salvation process as described in the KRM exactly match with the nitrogenous bases in DNA and RNA. At this stage, I appreciate how this may seem contrived, but I am going to show a lot of evidence to support it, some of which is very strange.

**The metaphor key**

I suggest that the different locations and actions of all nucleotides represent subtly different 'packages of Holy Spirit activity': the

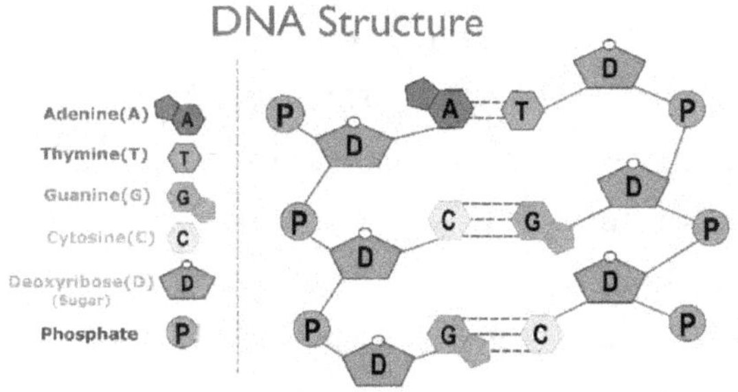


**Figure 2.2: Three base pairs; grayjay/Shutterstock.com**

nucleotides in one strand of DNA represent Holy Spirit activity with the Bible or any of the Trinity God; and in the other strand, His activity with Man. I will show how nucleotides in (single-stranded) RNA represent Man after salvation, and I will show that RNA has significant similarity to Genesis 2 in its slanting towards Christ. All enzymes are (by L-chirality of their amino acids) representing work accomplished by Jesus Christ, and at the same time, proteins in general may also by L-chirality parallel producing a Christ-like character; and this is all achieved by a breaking down and rebuilding process (biblically, you could say, *'katabole kosmos'* work). The proteins that these processes make (or synthesise) will represent the changes to saved Man's spirit or 'character'.

So, within this metaphor, the nucleotide itself is always representing a light bringer (phosphate), the Holy Spirit (by the D-chirality of ribose) and one of: a man, Jesus, the Holy Spirit Himself and the Word of God, depending on which base is attached. (Please note that 'light' is the word God uses in Genesis to define entities from a goodness aspect; Jesus, God the Father, the Holy Spirit, the Bible and saved Man may all be considered 'light givers' or light 'sources' and are described as such in the KRM; they can give '*light*' – 'images' giving 'likenesses'.)

DNA is a bit like a ladder, but the sides are twisted to form the famous double helix. The 'rungs' of the ladder are two nucleotides that join (by hydrogen bonds) to each other at their nitrogenous base 'ends' to form a 'base pair'. For this metaphor a 'rung' can be considered to be like a 'line' of six compounds: phosphate, ribose, nitrogenous base, nitrogenous base, ribose, phosphate. This makes for an instant comparison to the Genesis 'before-and-after-salvation' sentence structure and also, very specifically, it is in the same 'two sets of three' format just like Genesis 1:11-12. Rather than the two sentences in Genesis, we have two 'nucleotide sentences' linked together as a base pair; and I suggest **that <u>all</u> joined nucleotides tell the story of salvation from a beginning to an endpoint just like Genesis 1:11-12.**

This similarity to Genesis is established with different allocations to the strands: one from God, one in Man. Pre-salvation (the first DNA nucleotide in the 'rung'): God sends forth light, the Holy Spirit works for God on these 'lights' and He presents to Man the 'indwelling package' contained in the different nitrogenous bases. After salvation (the second DNA nucleotide in the base pair or 'rung'): the 'indwelling package' is now in Man; the Holy Spirit now works with it in Man, and makes a new 'light' in Man ('*the light of life*' – John 1:4[12]). Thus, with my allocations, DNA seems to share the same 'hidden design' as Genesis 1 and 2 by containing a before-and-after-salvation comparison

[12] John 1:4 (KJV): ***In him was life; and the life was the light of men.***

by using a ‘two sets of three’ pattern. But is this just a result of my ‘random’ allocations, are these parallels justified?

### A dynamic metaphor: other similarities

As we are using living processes, this ‘nucleotide metaphor’ presents as more of a dynamic, not just a start and end position description – a timeline; so the changes to get from one to the other become significant and also metaphoric. Please note that DNA is a polymer – a macromolecule – and the (complementarily paired) two nucleotides or ‘base pair’ are the monomers. The bond between monomers is not to an identical part of each monomer; rather it attaches from the phosphate group (‘light giver’ by ‘name’) to the sugar group (the Holy Spirit by chirality) and then from a different part of the sugar group and back to the next phosphate group. So, by this difference, when following the bonds’, and moving **between sequential base pairs**, you can go in different directions: from the working Holy Spirit to light, or light to the working Holy Spirit (from an ‘individual nucleotide point of view’). The bonds are at sites called ‘primes’ – named after which carbon atom has the bond: three (ribose end) or five (phosphate end).

Once this is understood...

### DNA replication is an overview of the salvation process

Throughout Genesis 1 and 2, the KRM describes the salvation process and how Man should evangelise and spiritually grow after that. I would expect this ‘nucleotide metaphor’ to do exactly the same.

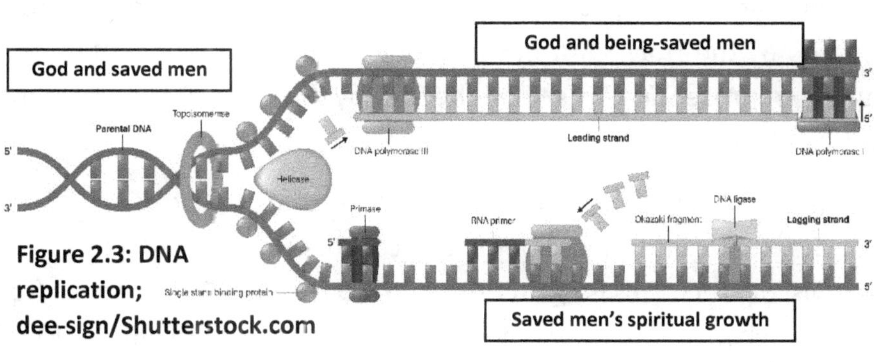


**Figure 2.3: DNA replication; dee-sign/Shutterstock.com**

The DNA replication process reveals salvation and spiritual growth. At replication, DNA splits into two single strands, still the same: one representing God and one saved Man in the 'nucleotide metaphor'. For the 'God strand' or 'leading' strand, the process moves to parallel the mechanics of achieving salvation itself in 'other' men, and we know saved Man is not a part of these mechanics and so his 'strand' is out of the way. God replicates His 'leading strand' to get to a two-stranded DNA again, and the new strand represents what the Holy Spirit 'copies' or uses from God (all that is metaphorised in the nitrogenous bases) and gives them to add to or change the to-be-saved Man's spirit. (Please note, this is still all about changes to Man's spirit and not his body, and not making his whole spirit from scratch.)

There are very subtle similarities in this biological process to salvation by Christ. The 'God strand' is replicated by various protein enzymes (metaphorising Jesus' work by chirality), and one of these enzymes forms the new DNA strand (metaphorising changes given to the to-be-born-again Man's spirit). It does this by reading the 'God strand' prime 3 to prime 5 (metaphorising the Holy Spirit looking to God's 'light'). But it forms the 'Man strand' (Man's reborn spirit) prime 5 to prime 3, and this is consistent with the indwelling gifts then 'downloading' from God to the Holy Spirit (the opposite direction) and then to Man. The God strand does not change (is conserved) and the newly born-again spirit has new additions. The replication process starts with an enzyme called 'primase' – a protein that makes an RNA 'primer', and this is consistent with the salvation process being initiated by the Holy Spirit but made by Jesus' work (the protein by L-chirality representing Christ and the RNA by the D-chirality of the ribose element in RNA representing the Holy Spirit).

The Man strand or 'lagging' strand also re-forms itself but incrementally. These increments are called 'Okazaki fragments' after their discoverers. This is again from prime 5 to prime 3, and hence from God to Man, as is expected in that this is representing Man's incremental spiritual growth. It also begins with primase (the Holy Spirit by Jesus's works) The different nitrogenous bases between the strands are again symbolising specific aspects of spiritual growth.

Below, Man's nucleotide ownership 'letter' is mentioned first to reveal the complementary function and nature of some of the interactions in saved Man's 'base pairs': the indwelt Holy Spirit interpreting the Bible for Man (GC); the living word (the Bible growing in Man through the Holy Spirit (CG)); Jesus in Man leading to Christ-like behaviour and fellowship with other saved men – '*body of Christ*' (AT); and Jesus as a Man leading to Jesus as God in Man (TA). Wave after wave of nucleotides maturing saved Man spiritually.

This is a neat summary of Man's spiritual growth. Thus, spiritual growth and evangelisation (in DNA replication) are closely linked, and this is consistent with general KRM structure as found throughout Genesis 1 and 2. (Even 'Okazaki' has a literal meaning of 'hill cape' – 'a high body of land extending into waters' – and this is reminiscent of the '*earth*' and '*waters*' metaphorising spiritual growth in the KRM.) It may also reveal meaning in Matthew 5:14[13], where the use of '*hill*' may be more significant than thought.

**RNA protein formation metaphorises making Man 'Christ-like' – the '*body*'[14] becomes more and more a suitable '*bride*' for Christ.**

The transcription (copying) of DNA to RNA and the subsequent work of RNA in making proteins (called translation) reveal Jesus Christ. Please note that the RNA events move to 'after salvation' (after DNA is made in the metaphor). In consistency with this, RNA does not have

---

[13] Matthew 5:14 (KJV): ***Ye are the light of the world. A city that is set on an hill cannot be hid.***

[14] The mysterious '*body of Christ*' comprises all Christians and their activity, with Christ as the head of the body, and it centres on Christ's image life given to Man. The '*bride' of Christ* is a personal relationship between Jesus, the Holy Spirit and the individual saved Man and involves Man's whole spirit; both are classed as 'image life' in the KRM in Genesis, the '*male and female*' of Genesis 1:27.

thymine in it: thymine is replaced by uracil (U) in RNA, and the other nitrogenous bases stay the same. From a mainly 'image' point of view, at salvation, the Holy Spirit places a part of Jesus' spirit, a part of Himself and the Word of God (Bible) in Man, but He also regenerates Man's spirit (Titus 3:5[15]) – uracil. Please note how uracil is 'closely related' (or paired with) adenine (Jesus Christ); this is consistent with saved Man's relationship to Christ – being in Christ and Christ in him. The DNA parallel above does not need to contain uracil as 'Man' (thymine) is a broader term which may mean Jesus as a Man being presented (before or after salvation), saved Man, or even unsaved Man (just as *adam* as opposed to *ha-adam* is used in Genesis in the KRM).

The established named functions of RNA (messenger RNA, transfer RNA and synthesising ribosomal RNA) are consistent with the Holy Spirit's known functions: approaching Man and then giving and making the indwelling at salvation. The transcripted RNA (carrying aspects of the 'copied' indwelling package from the 'God strand') goes on to form proteins, and the proteins formed contain left-handed or L-amino acids (by chirality – of Christ). This will be covered in much more detail in Chapter 13, but proteins, generally, can be considered as representing Man being in Christ (*'the body'*) and then Man becoming more and more Christ- and Holy Spirit-like (*'a bride'* of Christ).

**I re-emphasise that RNA and protein synthesis, in <u>all of creation,</u> testify to the Lord Jesus Christ – a Christ-themed creation.**

But that is not all: other nucleotides are used to make energy. Energy is needed for the functioning of all cells both internally and for 'export'. An end product of the DNA/RNA/enzymes working together in the mitochondria is the universal energy supplier – ATP. Most do not realise that this too is a nucleotide very similar to the adenine-based

---

[15] Titus 3:5 (KJV): ***Not by works of righteousness which we have done, but according to his mercy he saved us, by the washing of regeneration, and renewing of the Holy Ghost;***

nucleotide in DNA (see Figure 4.1). But it contains **three** phosphorous components (three 'light bringer' 'packages') as well as the similar ribose (the Holy Spirit by chirality) and the same adenine (representing the Lord Jesus Christ by etymology). This molecule can be 'interpreted' to mean that all 'energy outputs' of ATP (that will be fuelling processes that metaphorise an aspect of salvation) can be considered to be 'works' that are the result of Jesus (adenine) and the Holy Spirit (ribose) working together. The energy for DNA replication comes from base nucleotides in this triphosphate state as they can all exist in this state. In each case, they lose phosphate groups, and this gives the energy for replication and transcription. This makes the phosphate groups or 'light bringers' the 'drivers' or the 'energy currency' for the whole salvation process.

It should be noted that the root word meaning of 'nucleus', from which 'nucleotide' derives, is '**seed**' or 'kernel', and this seems apt in metaphorising aspects of salvation or the Bible itself, which are also referred to as '*seed*'. Thus, *nucleo-tide* may metaphorically mean a '**part of the Bible**'. It may also be, for this metaphor, that as codons (the base unit of information in DNA) are always made of three nucleotides, these may parallel 'light' as ultimately coming from three varying 'sources' (this is also covered in Chapter 13). Even the presence of three hydrogen bonds between G and C, and only two between A and T, may signify the number of participants in the interaction – the Holy Spirit, the Bible and Man vs Jesus and Man respectively. I suggest that as the release of energy from ATP involves the addition of water, and as, biochemically, water seems to be always present in the vicinity of DNA, this makes it reminiscent of the '*living water*' of Genesis 2; and also with '*waters*' acting as a metaphor for the Bible.[16]

I additionally point out that DNA in mitochondria is not in strands (as is found in the nucleus of all eukaryotes) but takes the form of a

---

[16] Ephesians 5:26 (KJV): ***That he might sanctify and cleanse it with the washing of water by the word,***

circle (a ring); and, as the 'likenesses' or energy it synthesises are from Jesus, the Holy Spirit and the Bible plus or minus Man's regenerated spirit, I wonder if 'rings' signify 'completeness' or the 'family of God' too. All nucleotides consist of heterocyclic rings (picture pentagons or hexagons with carbon or nitrogen at their vertices); purines have two rings joined at one of their 'sides', and I wonder if one represents Man or the Bible (six-membered), and one God (five-membered) with the border their shared aspects. Finally, where and how the elements bond may need consideration too, because, amazingly, here comes...

## **The Proof: Eight Chemical Formulae** (hidden in plain sight)

Now it is going to get strange, and also profoundly provocative! Consider this: I suggest God does not do metaphors half-heartedly; if we look at the different chemical formulae of the ribose element in RNA and DNA ($C_5H_{10}O_5$ and $C_5H_{10}O_4$ respectively), we see that the difference is just one more oxygen atom in RNA. I will show how in this 'nucleotide metaphor' the main difference between the functions of DNA and RNA is the extra 'work' RNA has (as it is long-term after salvation) in working with the extra 'light' of saved Man's reborn spirit continually being worked on to become more 'Christ-like' (Romans 8:29[17]); this is the light that makes Man *'the light of the world'*. Does the extra oxygen atom represent this extra 'light' activity?

The centre of the phosphate group ('$PO_4$') is phosphorous ('P') – 'the light bringer'. Logically, does this make the 'O' the 'lights'? If we look at the root word meaning of 'oxygen' – it means 'sharp' or 'acidic', and 'producer' ('gen' meaning the producer) – then oxygen has a meaning of 'acid producer'. This link between oxygen and acid production will reveal much, perhaps the most important aspect of the whole process.

---

[17] Romans 8:29 (KJV): ***For whom he did foreknow, he also did predestinate to be conformed to the image of his Son, that he might be the firstborn among many brethren.***

If we look at 'ribose', it is associated with 'acid' too (from 'saure') and is also related to 'arabinose', which has a root word meaning of the 'gum of an acacia tree' – 'acacia' also has an etymology of meaning both 'without evil' and a 'prickly tree'. (The Holy Spirit is metaphorised as *'a tree'* in several places in Genesis, notably *bdellium* ('the fragrant resin of tree') in Genesis 2:12 and a *'fruit tree'* in Genesis 1:11-12; and in my opinion the *'pricks'* in Acts 9:5[18] are the promptings of the Holy Spirit too.) As He is certainly without evil, this makes for a very comprehensive summary of His roles being hidden in the etymology of *'ribose'*. Is God using the elements themselves to consistently signify something? In the case of oxygen, does the 'light' (that makes acid) produce acid for the Holy Spirit (the 'acidic one' or 'acid tree') for Him to work with? I suggest, startlingly, God repeatedly uses all the component elements in nucleotides to extend His metaphor.

Throughout the KRM in Genesis 1 and 2, the 'interactions' that Man, Jesus, the Holy Spirit and the Bible have in the salvation process are 'characterised' by God as involving *'images'*, *'likenesses'*, *'living water'*, *'lights'* and *'trees'*. There is an overlap within these categories, but these are God's own 'divisions' in Genesis 1 and 2. At the same time, I am about to show how the elements that make up all of the components in DNA and RNA (carbon, nitrogen, hydrogen and oxygen) have root-word meanings in their names that give a **precise 'crossmatch'** – the element names and their root-word meanings match with God's categories in Genesis. Please note, as above, oxygen (*'light'*) is via 'acid' and acidity is often considered as giving hydrogen ions – 'acid' seems to be the unit or 'currency' of *'light'*. Not only that but, I suggest, the number of 'cross-matched' interactions in Genesis always reveals the element count in the formula of the components they constitute. Thus, deoxyribose and ribose in DNA and RNA have

[18] Acts 9:5 (KJV): ***And he said, Who art thou, Lord? And the Lord said, I am Jesus whom thou persecutest: it is hard for thee to kick against the pricks.***

four and five oxygen atoms respectively, as they are representing the **working** Holy Spirit in action; and He interacts with four or five 'light sources' in Genesis 1 and 2, and chemical formulae reflect this. Genesis 1 and 2 will describe these interactions in great detail and from multiple perspectives too; and it will be relatively easy to count them there (often from one verse). Hence **deoxy**ribonucleic acid represents working with 'light' (or 'oxygens that induce a Holy Spirit response') from four light sources (God the Father, Jesus Christ, the Holy Spirit and the Bible) – '$O_4$'.

After salvation (still in DNA but the other strand now – the 'man' strand), Jesus through the Holy Spirit makes the two *'great lights'* – **in Man** on Genesis Day 4; and the Holy Spirit and the Bible also act like *'lights'* **in Man** after salvation and beyond (two of the trees in Genesis 2:9 and I will come to trees and lights being synonymous shortly). That makes four *'lights'* again, but now in Man – a different four now.

In the long term, after salvation, now in RNA in the metaphor, Jesus puts in another appearance in that the regenerated aspect ('uracil') of saved Man is increased and Man's original spirit is continually changed by the Holy Spirit bringing more of Jesus' *'light'* to carry out the regeneration; and this makes it 'brighter' for Man, enabling him to become more Christ-like. Long-term after salvation, the Holy Spirit, thus, works with the four lights already in Man and now the extra light of Jesus Christ; five light sources – the ribose part of RNA must have '$O_5$'. Christ's 'light' is on the 'outside' and placed inside by the Holy Spirit, and that too then shines 'through' and onto the 'outside' – *'light[s] of the world'* both (John 8:12[19]). (It is very provocative that DNA fluoresces, and most so when thymine and adenine are adjacent – Jesus and Man – the *'body of Christ'* being the light of the world?)

---

[19] John 8:12 (KJV): ***Then spake Jesus again unto them, saying, I am the light of the world: he that followeth me shall not walk in darkness, but shall have the light of life.***

The 'P' in the first phosphate group ($PO_4$) or 'God strand of DNA' is representing the authority of God the Father, commanding: *'Let there be light'* in Genesis 1; sending out four 'lights'. The 'P' or 'light bringer' in the second or 'Man strand' is Jesus Christ as the *'light of life'*, and in RNA the 'light bringer' includes the *'body of Christ'* (all Christians). Phosphate groups ($PO_4$) represent the authority to 'bring light' for salvation and will also be very significant in the biological KRM.

**The chronologically correct sequence** of events is apparent: the first nucleotide in a base pair of DNA is before salvation; the second of the complementary pair is at and after salvation; after that the two strands represent the ongoing relationship of Man and God together; and single-stranded RNA is in the long term after salvation.

**An element of truth**

Let us consider the etymology and root word meanings of the other elements that constitute nucleotides, my proposed 'cross-matching' of them to aspects of the salvation process, **and God's own equivalent categorisation.** I will introduce more of the metaphor-holding words in Genesis at this stage, and please remember that all of these are confirmed by adhering to some complicated structural rules in Genesis.

**Carbon:** has etymology associated with *'charcoal', 'coal' or 'ember', or a 'glowing ember'* (a 'flame in the middle') – this is consistent with carbon representing the spirit 'image' inside Man's body; **God's 'image life' given to Man – Man's image** (a definition of images will follow).

**Nitrogen:** has etymology associated with *'native soda forming'* – the 'native' can be Man, so, Man's own 'fizzing' identity, this is consistent with nitrogen representing Man's likenesses; **God's likenesses given to Man – Man's identity** (a definition of likenesses will follow).

**Hydrogen:** has a meaning of *'water forming'* and this meaning is consistent with representing Man producing 'living water'; **all interactions that act to produce *'living water'* in Man.**

**An inadvertent aside:** On a personal level, I particularly like inadvertent naming; and I will show that this happens with the Cenozoic era too. There is something very pleasing to me about root words (named by the secular) confirming biblical truth!

Now, I want to show just how elaborate the KRM is in Genesis and to also show, especially for those who go on to read the KRM book, just how refined the 'code' is. If carbon, hydrogen, nitrogen and oxygen represent image life, living waters, likenesses and light in Genesis, respectively, then how are they referred to in Genesis? What is their metaphor there? These metaphoric representations do not have to be memorised at this stage but they will all crop up later in this book (brackets indicate representation in the biblical KRM in Genesis):

**Image:** the 'substance' of the spirit (*'nephesh chayyah'*, *'cattle'*, *'light'* (source) and *'seed'*), or the whole spirit 'being' (*'earth'*, *'living creature'*, *'body of Christ'* (*'male'*), *'bride' of Christ* (*'female'*)), *'beast of the earth'* becoming '***the** beast of the earth'*, *'a regenerated spirit'*. (All saved Man's newly created images from God are listed in Genesis 1:27; and Genesis 1:24-25 lists all Man's and *'regenerated'* Man's own contributions – these will be the 'carbon' or 'image' verses.)

**Likeness:** skillsets, attributes, identities, attitudes and behaviours (*'light'*, *'fowl of the air'*, *'beast of the earth'*, *'fish of the sea'*, *'every living thing that moveth'*, growing *'trees'*, evangelising (*'multiply'*, *'fruitful'*) and spiritual growth (*'replenish'*, *'subdue'* and *'fruitful'* again)). All saved Man's after-salvation likenesses are listed in Genesis 1:28 – these will be the 'nitrogen' verses. (Man's original likenesses are called *'beast of the earth'*, and these become a part of the *'body of Christ'* after salvation and hence are no longer classed as likenesses.)

**Living water**: the result of an amalgamation of Man, the Bible, Jesus Christ and the Holy Spirit. The *'living water[s]'* are a separate entity to 'images', but do mingle with images (*'earth'* or *'eres'* – Genesis 2:11 and 13) to produce different attributes as represented by the names and destinations of four rivers in Eden. *'Living water'* is a bit of a mystery, but the Bible describes it as literal, and hence it is visible in the spiritual realm flowing out of Man. Genesis 2:10 describes how this *'living water'* forms – the 'waters verse' – more like an image statement; and Genesis 2:10-14 describes the function and properties of this *'living water'* – the 'living water function verses' – more like a likenesses statement. The nitrogenous bases of DNA metaphorise making Man, and so they relate to **forming** the *'living water'*, but the working Holy Spirit (ribose) will work with the **functioning** *'living water'*.

**Light**: a living entity from a 'goodness point of view'; it tells right from wrong, good from evil, signs and seasons, all truth; and in Jesus' case, His given 'image' life makes for a 'light' in Man – a change to Man's existing spirit image – and this change specifically makes 'lights' (*'two great lights'*). Days 1, 4 and 5 contain the 'light verses'.

Do the other atomic proportions within chemical formulae for nucleotides give 'cross-referenced' accurate counts for actions in the KRM-described salvation process? Well, yes, they do, and I can even say that once the rules are known, I can predict the chemical formulae of all components of DNA and RNA just from Genesis alone, or at least confirm them. This needs a pause: **I can deduce the chemical formula of the building blocks of all life just from Genesis 1 and 2!**

It is all by a simple cross-referencing with the Genesis text, and the 'image-then-likeness' structure to the KRM greatly facilitates this process. It is also important to note that this comparison is between chemical formulae and the **Genesis text itself** – with Genesis' own categories and perspectives, which may overlap in our eyes. In effect, the chemical formulae adhere to specific parts of Genesis 1 and 2.

**In the KRM:** Genesis 1 and 2 describe the salvation process multiple times (at least 15 times), and from multiple perspectives. Genesis 1:27 has the perspective of what God gives to Man that is a 'created image'.

Genesis 1:26-28 describe *'God'* (plural) declaring *'us'* (plural), making or creating all of these God-given 'image lives' and likenesses; this plural (*'us'*) includes Jesus Christ and the Holy Spirit. (In my opinion Genesis 1:27 is the highest text, for information, in the entire Bible.)

**The carbon (images) count from God to Man** (Genesis 1:26 (KJV))

***So God created man in his 'own image';*** Jesus Christ's image is given to Man (please note that Man's own spirit is remade by the Holy Spirit at this time too, but it does not class as a 'created image' from God here, as it is a 'remake' and originates from Man). Both Jesus and the Holy Spirit are involved in this process of creating of 'one image life' for Man's saved spirit.

***'In the image of God created he him';*** Jesus and the Holy Spirit are placed in Man's living waters; both Jesus and the Holy Spirit are involved in this process of giving 'two further image lives' to Man.

'***Male*** (the *'body'*) ***and female*** (the *'bride'*) ***created he them.***' As above, both Jesus and the Holy Spirit are involved in this long-term process of 'two more image lives' being given to saved Man's spirit. That makes for a count of five 'image' inputs each (into saved Man's spirit) for both the Holy Spirit and the Lord Jesus Christ. The role-representative nitrogenous bases of Jesus and the Holy Spirit, adenine and guanine, contain '$C_5$'.

**The nitrogen (likenesses) count from God to Man**

**Genesis 1:28** lists all of Man's likenesses: ***Be fruitful (1.*** *spiritually grow – especially Christ-like****), and multiply (2.*** *evangelise****); replenish the earth (3.*** *get more of Christ****) and subdue it (4.*** *Christ changing Man's original sinful spirit****) and have <u>dominion</u> over the fish of the sea (5), and over the fowl of the air (6), and over every living thing that moveth upon the earth (7.*** Holy-Spirit- and Christ-like maturation, and incorporating the Bible and other men***)*** – 'Man's seven likenesses'. I will later show how the salvation process descriptions are always divided into three categories: Christ- and regenerated-Man centred;

Holy Spirit and original spirit of Man centred; and finally, both Christ- and Holy Spirit-like maturation. The Christ centring is all just given at salvation but the rest have to be grown (the use of *'dominion'* shows the onset of these growing likenesses). The Holy Spirit is the work force for all of salvation and I will later show that by the use of *'bdellium'* His presence is noted. The representative nitrogenous bases of Jesus Christ and the Holy Spirit, guanine and adenine, contain '$N_5$' (for Christ it is 1, 2, 3, 4 and 7, and for the Holy Spirit's it is 3, 4, 5, 6 and 7).

**The hydrogen (living water) count from God to Man**

***Genesis 2:10 (KJV)*** describes six interactions of a river: ***And a river went out of Eden*** (1) ***to water the garden*** (2)*;* ***and from thence it was parted, and became into four heads*** (3, 4, 5 and 6), with each river then leading to various 'likenesses'. The KRM shows how the *'garden'* metaphorises the indwelling of Jesus and the Holy Spirit and the 'river' metaphorises Man's *'living water'*. The river interacts (*'water'*) with the garden and then goes on to form the *'four heads'*. Thus, *'out of Eden'* is upstream of Jesus and the Holy Spirit, and they do not participate there. Please note again, this sentence is about actions that form *'water[s]'* and not the functions of the 'formed' water(s). Therefore, Jesus and the Holy Spirit partake in five of the six interactions that form *'living water'*; the representative nitrogenous bases of Jesus Christ and the Holy Spirit, adenine and guanine contain '$H_5$'.

**The oxygen (lights) count from God to Man and then in Man**

*'Two great lights'* are mentioned in Day 4, and the KRM reveals these are Man's spirit being added to and changed. The Holy Spirit and the Bible living in Man also fulfil the description of what 'lights' should do.

Jesus gives saved Man 'image life', and this 'image life' gives to Man the *'light of life'*; as confirmed in: ***'the life*** *(*'image life'*)* ***was the light of men'*** (John 1:4); the light is now Man's and not Jesus'. Jesus gives saved Man this light with His image life (not light itself) via the Holy Spirit for Man to gain this *'light of the world'* status (which is also revealed to other men by the Holy Spirit). I show in the KRM book how, in Genesis 2:9[20], *'trees'* are representing the Bible and the Holy Spirit – the *'tree of life'* is the Bible (Bible and Holy Spirit together) and the *'tree of knowledge of good and evil'* is the Holy Spirit alone (scripture confirms the Bible is 'living'; Hebrews 4:12[21].)

I suggest again that **the term *'tree'* is a way of describing 'Holy Spirit-aided functioning and growing lights'** and this explains why the root word for 'ribose' (the functioning light) is also based on a tree. The use of *'tree'* in Genesis 1:11-12 and 29 is consistent with this, and I will show that the use of *'trees'* is a reoccurring way of recognising the intermediary Holy Spirit giving and 'nurturing' 'light' in Man.

The Christ-centred categories act with more and more of Christ's light, on the *'greater'* and *'lesser lights'* in Man; and the Holy Spirit-centred categories act with more of the Holy Spirit's image life and likenesses, but with the Holy Spirit and the Bible acting as lights in Man. The Christ- centred are just given and are now Man's, but the Holy Spirit is a more a 'permanent tenant'. This splitting is very significant.

Hence, for chemical formulae, with 'O' representing **giving *'lights'* to saved-Man's** spirit: Jesus (adenine) will **not** have an 'O' (He gave 'spirit life' only and this, in turn, gave light); the Holy Spirit (guanine) and the Bible (cytosine) will have one – 'O'; Man's original spirit (thymine) will be involved with (or house) two lights – '$O_2$' (the Holy Spirit and the Bible); and Man's regenerated spirit (uracil) will be involved with (or house) the *'greater'* and *'lesser light'* – also '$O_2$'.

---

[20] Genesis 2:9 (KJV): ***...made the Lord God to grow every tree that is pleasant to the sight, good for food, the tree of life also in the midst of the garden and the tree of knowledge of good and evil.***

[21] Hebrews 4:12 (KJV): ***For the word of God is quick*** *(can also mean' living'),* ***and powerful, and sharper than any two edged sword, piercing even to the dividing asunder of soul and spirit, and of the joints and marrow, and is a discerner of the thoughts and intents of the heart.***

The etymology and chemical formula in nucleotides are always all about what an entity **contributes** to an end point in Man's spiritual condition or status – an itemised declaration of the origins.

- **Adenine ($C_5H_5N_5$) represents the Lord Jesus Christ, and its chemical formula reveals what He provides for saved Man:**

As previously considered, the Lord Jesus Christ contributes: '$C_5$'.
As previously considered, the Lord Jesus Christ contributes: '$H_5$'.
As previously considered, the Lord Jesus Christ contributes: '$N_5$'.

The chemical formula for adenine is present in Genesis and, in a way, this confirms its metaphoric representation of Jesus Christ.

- **Guanine ($C_5H_5N_5O$) represents the Holy Spirit, and its chemical formula reveals what He provides for saved Man:**

As previously considered, the Holy Spirit contributes: '$C_5$'.
As previously considered, the Holy Spirit contributes: '$H_5$'.
As previously considered, the Holy Spirit contributes: '$N_5$'.
The Holy Spirit is a 'light' in saved Man: 'O'.

The chemical formula for guanine is present in Genesis and, in a way, this confirms its metaphoric representation of the Holy Spirit.

The chemical formulae of these two nitrogenous bases appear to be driven by biblical truths; can this all just be contrived or a 'stretching' of the biblical text? Do the same rules apply (**with no changes in mechanism and metaphoric meanings**) to all of the other nucleotides and components of DNA and RNA? That makes for eight compounds in total, and it seems extremely unlikely that this could happen randomly.

There are two representations for Man's spirit: 'thymine' representing his original or whole spirit, or 'uracil' representing just the regenerated **part** of his old spirit – his 'uracil' was made '*from the dust of the ground*' (Genesis 2:7) or from his original spirit ('thymine'). This means Man's regenerated spirit has its origin in Man's original spirit.

**3. Thymine represents Man's original spirit, and its chemical formula reveals what it provides for saved Man:**

Man's original spirit (thymine) provides: (1) unchanged aspects of his own original spirit (*'earth'*, *'ground'* or *'the living creature'* of Genesis 1:24); (2) the constituents for his regenerated spirit (*'dust of the ground'* – Genesis 2:7); (3) Man's own *'beast of the earth'* are his likenesses but they are transformed to the image of Christ and this is confirmed with the addition of the definite article *'**the** beast of the earth'* – a 'likeness' becoming an 'image' (Genesis 1:24-25); Man's original spirit is also a part of 'the collectives': (4) the *'body of Christ'*; (5) and *'a bride'* of Christ (Genesis 1:27) – five inputs – '$C_5$'.

'Waters' (*'river'*) are originally Man's own waters (he is 'Eden' in the KRM), so he is involved with *'a river went out of Eden'*, *'water the garden'* and *'became into four heads'* (six in total) – '$H_6$'.

Man's seven likenesses in Genesis 1:28 have two 'threads' and an end game, and although they all remain Man's identity (likenesses), some do not originate in Man. Man's original spirit (his 'thymine') is mentioned (as *'land'* or *eres*) in the category-matching passage of Genesis 2:11-13 (below the category-matching term is in brackets within the wider brackets), and so, these will reveal Man as an origin for some likenesses: N2 (*'multiply'* or (*'Havilah'*)) and N4 (*'subdue'* or (*'Ethiopia'*)) – thus, thymine has two 'nitrogens' – '$N_2$'.

Again, as previously described, the original spirit of Man deals with (or houses) the lights of the Holy Spirit and the Bible – '$O_2$'.

Genesis reveals that the chemical formula of thymine is $\mathbf{C_5H_6N_2O_2}$, and this, in a way, also confirms its validity as a metaphor for Man.

4. **Uracil represents the regenerated aspect of Man's spirit, and its chemical formula reveals what it provides for saved Man:**

Man's regenerated spirit's participation is also revealed in Genesis 2:11-13 and mentioned twice as *'onyx stone'* ('good gold', *'bdellium'* and *'onyx stone'* are also included by the use of 'the same' in Verse 13). Man's regenerated spirit has direct interaction with the *'greater'* and *'lesser light'* – '$O_2$'. This *'lesser light'* is continually being regenerated.

The regenerated aspect of the spirit does not contain all the initial components of it: it does not contain the whole original spirit (*'earth'* and *'living creature'* are not the same as the *'ground'*, the ground is a subset of *'earth'* in Genesis). 'Uracil' (the regenerated spirit) itself is a part of the saved Man's spirit and also contributes to *'beast of the earth'* (Genesis 2:20), and is a part of the *'body'* and a *'bride'* of Christ (the regenerated man is one of the 'cattle' of Genesis 2:20-25) – '$C_4$'.

As the term *'cattle'* can also represent Man's regenerated spirit being 'in' or under 'God's shelter' (in the *'body'* and *'bride'*), and as *'earth'* can be flipped to mean Man's spirit is doing the *'bring[ing] forth'* in Genesis 1:24-25, Genesis 1:24 becomes a carbon count for thymine, and Genesis 1:25 becomes a carbon count for uracil. (Genesis 1:25 is going to be very, very significant in its very specific ordering.)

For *'living water'*: the regenerated spirit of Man is placed *'in the garden'* (Genesis 2:8), and therefore this can be considered as 'after' the garden has been watered, as the *'living water'* *'went out of Eden'* and ***'thence'*** *'water[ed]'* the *'garden'*; saved Man is not a part of those first two interactions – just the four *'heads'* of the rivers, hence '$H_4$'.

The chemical formula for uracil is confirmed in Genesis as $\mathbf{C_4H_4N_2O_2}$, and this, in a way, confirms its metaphoric representation as the regenerated spirit of Man.

5. **Cytosine represents the Word of God (the Bible), and its chemical formula reveals what it provides for saved Man:**

Day 5 and Genesis 2:9-18 are 'the Bible verses'. Perhaps it is 'necessary' to know the Word of God to receive the associated gifts.

The Bible is represented by *'waters'* throughout the KRM, and what the *'the waters'* of Day 5 *'bring forth'* can be, in part, attributable to the Bible and these include Christ's image life, the Holy Spirit's image life, and a part (by 'filling' the *'seas'* and *'earth'* of Verse 22) in making a *'bride'* and *'body of Christ'* and – '$C_4$'.

The Bible is a part of the *'tree of life also in the midst of the garden'* (Genesis 2:9); hence the Bible has the same number of interactions in making *'living water'* flow, as do Jesus and the Holy Spirit – '$H_5$'.

The Bible has its part in the Holy Spirit centred, and long-term declarations of Genesis 1:28: they are *'fish of the sea'*, *'fowl of the air'*, and *'every moving thing that liveth'* (N5, N6 and N7) – '$N_3$'.

The Bible is a *'living word'* in Man and is classed as a *'light'* throughout this metaphor and elsewhere (Psalm 119:105) – 'O'.

The chemical formula for cytosine is confirmed in Genesis as **$C_4H_5N_3O$**, and this, in a way, confirms its metaphoric representation as the Word of God or Bible.

6. **Deoxyribose (DNA) before salvation, at and after salvation and (7.) Ribose (RNA) long-term after salvation** – all by the **working** Holy Spirit; **(8.) Phosphate ($PO_4$):**

The change from '$O_4$' to '$O_5$' in ribose and the total number of 'O's in the phosphate group have already been discussed.

The Holy Spirit's representation by chirality as ribose is different to nitrogenous bases, and is recognising His works – the Holy Spirit actually **making then using** likenesses (referred to as *'yielding'*, *'giving light'*, *'moving'* and *'creeping'* in the KRM). He can present Himself: the working Holy Spirt (ribose) presenting the image-and-likeness-and-light-giving Holy Spirit (guanine) – 'the fertile product of the air'.

The chemical formula of ribose will show what the Holy Spirit is working **with, on or through**, as opposed to nucleotides revealing what they actually give or contribute in a specific etymologically described role ('a fertile product again'). This will subtly change within the before, after and long-term-after timeline that paired nucleotides and RNA represent. It sometimes seems that 'ribose' is using some of Man's indwelt spirit '$C_5$' gifts in the 'before' metaphor in DNA (before they have been given); I suggest this is referring to other saved men, and this will be something that repeats a lot – the Holy Spirit working before or after salvation with a saved Man and also with other saved men helping in this process.

The river names are metaphorically significant and will be discussed in great detail later, but to summarise: we have four river names and three place names, and one of these place names ('*Havilah*') is described as having three resources. Each of these entities will represent a likeness – 'the ten water likenesses'. These too, will be very significant in the rest of this book. If the working Holy Spirit (ribose) is using or making the gifts '*living waters*' give, then that is '$H_{10}$'.

The timeline becomes apparent: in the 'God strand' nucleotide; the Holy Spirit works and uses the four 'lights' (from God) '$O_4$'; He works with other saved men (working through them with their 'new image life' – '$C_5$' and using their 'living water' functional gifts '$H_{10}$' to present the indwelling 'package' to Man. In the second nucleotide of the base pair, there is a very specific change: the indwelling package is now in Man; 'image life' ('$C_5$') is made (and then used) by the Holy Spirit; '*living water*' being made and used by the Holy Spirit ('$H_{10}$'); and four '*lights*' in Man ('$O_4$' also being made and used, but a different $O_4$ now). The long-term after-salvation RNA ribose element (the Holy Spirit) changes again and still uses Man's indwelling image gifts ('$C_5$' and '$H_{10}$'), and also his same four lights, but He specifically works on the regenerated Man to make him more Christ-like from more of Christ's light. All of these together are used to make Man spiritually grow and to evangelise; and this long-term action of gaining more of Christ's light, constitutes an extra light ('$O_5$') for the Holy Spirit (ribose) to work with. There are subtle changes to the 'C' and 'H' counts to be discussed in more detail later. Although Man's likenesses – '*fruitful*', '*multiply*', *'replenish'*, *'subdue'*, *'fish of the sea'*, *'fowl of the air'*, and the long-term *'every living creature that moves'* – have contributions from the Holy Spirit, they remain under Man's, at least, partial dominion (Genesis 1:26-28), and hence they are Man's likenesses and, when used, Man's works from Man's perspective – there is no 'N' in ribose.

The chemical formulae for ribose in DNA and RNA are confirmed in Genesis as **$C_5H_{10}O_4$ and $C_5H_{10}O_5$**, and this, in a way, confirms their metaphoric representations as the **working Holy Spirit.**

A good level of understanding of the KRM is needed to appreciate the chemical formula aspect, and I think, as this is all metaphorical, it may be that (rather like parables: Matthew 13:11[22]) the Holy Spirit is needed to interpret the Bible in the first place before the KRM can even be recognised. This may diminish its use as a 'proof of God' for the secular, but I believe it is self-evident that this cannot be coincidental.

I am in head-shaking awe every time I consider this: God is not just picking the elements to fit His second narrative (add another 'O' here, take off an 'N' there!), as the compounds must have certain chemical properties for their very intricate, staggeringly complex and precise function – and these properties would change with such additions.

I conclude that there are really only two options: there is only 'one way' that life can be created and this would necessitate that the physical world itself and all the elements within it, all languages and all life, were created, made and formed with Jesus Christ's works a forethought – incredible; or it may be that there are other ways of 'creating the biochemistry of life', perhaps more streamlined than at present, but God chose this particular one in order to tell the story of salvation by Christ. Either way, **wow, wow and wow again!**

---

[22] Matthew 13:11 (KJV): ***... Because it is given unto you to know the mysteries of the kingdom of heaven, but to them it is not given.***

## The Pattern: Image to Likeness

As we found in Genesis, we would expect to see an 'image-to-likeness' structure throughout. The phosphate, adenine, guanine, thymine, and uracil are 'images', and their number of actions or the result of their likenesses are revealed in their chemical formulae when written in the standard Hill system (Hill system: first carbon, then hydrogen, and then the next elements alphabetically (CHNO). Well, carbon is the image, hydrogen is both image and likeness, and nitrogen and oxygen could both be interpreted as likenesses.)

The characterisation of 'likeness actions' is also in the etymological meaning (often 'roles') of the names of the compounds and elements. By this extraordinary method, the 'image-to-likeness' structure is maintained. Hence, in effect, each name has an 'image-to-likeness' structure built in.

There is also a constant 'image-to-likeness' change in the nucleotides themselves: phosphate (images), ribose (likeness) and nitrogenous bases (images); then the coupling action (likeness), nitrogenous base (image), ribose (likeness) and phosphate (image). The 'image-to-likeness' structure occurs at least 90 times in Genesis 1 and 2, and is a part of what is 'a-must-be-designed structure'.

### A three-day reverence to Christ

Nucleotides do not read left to right like writing, as they may be at any angle in the body. DNA has a twist that completes a circle about every ten nucleotides. You can look at one nucleotide from the phosphate end and then a nucleotide above or below it also from the phosphate group end; or you could read nucleotides together as one long 'base-pair' line. Both readings when considered from a metaphoric viewpoint have a built-in-three-days reverence to Christ:

**As a one-line base pair: (God)** Phosphate (1); ribose (2); base (3); **(Man)** base (4); ribose (5); phosphate (6).

(1) God shines light – three 'days' later the indwellings are in Man (4).

(2) The Holy Spirit works for God – three 'days' later in Man (5).

(3) The Holy Spirit presents the indwelling package – three 'days' later the *'light of life'* is in Man (6).

Reading both nucleotides from the 'light' or 'phosphate' end reveals a very close alignment with the six days of Genesis and also with John 3:3-5: phosphate (1); ribose (2); base (3); phosphate (4); ribose (5); base (6).

(1) Gods sends out light – three 'days' later Christ's light is in Man (4) – Day 1 and Day 4 (*'born-again* days').

(2) The Holy Spirit works for God – three 'days' later He works in Man (5) – Day 2 and Day 5 (*'born of water* days').

(3) The Holy Spirit presents the indwelling to Man – three 'days' later the indwelling is in Man (6) – Days 3 and 6 (*'born of the spirit* days').

**A repeating theme**

I concede that this whole comparison process is certainly peculiar, but the same design principles are now demonstrably in Genesis and DNA – the same underlying building blocks are used, and these building blocks tell the story of salvation by the Lord Jesus Christ. Well, well – who would have thought it? DNA actually testifies to the Lord Jesus Christ, I suggest addition of brackets to commonly used acronyms: DNA to **DNA (God and Man)** and RNA to **RNA (Jesus Christ)!**

Is this credible proof even on its own that there is a God and He is the God of the Bible? I believe it is a huge step in that direction – **Genesis and DNA must share the same author; 'DNA is the gospel truth'; and Watson and Crick did not even know *'the half'* of it!** Let's see some more examples...

# Chapter 3

# Porphyrins and the Purple-robed Lord Jesus Christ (I)

*A pigmented biochemical gospel in a universal metaphor*

## The Theory

I am going to consider whether the KRM exists elsewhere in biochemistry and physiology. First, I will consider the coloured 'porphyrins'; then some more nucleotides; then I will look in detail at 'electrons' and 'redox' reactions; and finally, I will consider one very specific biblical parable. In each case I will posit a parallel or metaphoric meaning based on the etymology of the compound's name – and these will, just as with DNA, all be 'biblically logical'.

I will ask you to bear with me for this process, because the 'etymological history' and their biblical parallels will not validate themselves until I show you how they come together to reveal that photosynthesis, cell respiration and protein synthesis all, astoundingly, adhere closely to the Genesis text – **the majestic Genesis text**. (This includes light-dependent and non-light-dependent reactions of photosynthesis, photorespiration, glycolysis, the Krebs cycle, protein synthesis and the electron transport chain.) I chose these because, apart from DNA biochemistry, they are perhaps the 'biggest' or most well-known biochemical processes.

'*Qara*' was the key to unlocking Genesis, and homochirality the key to unlocking DNA – both processes testifying to the works of the Lord Jesus Christ in astonishing detail. I suggest **all** biochemistry **may** be 'themed' by God to testify to Christ in extraordinary ways (or, more specifically, to adhere to the Genesis text in the way it also testifies to Christ). I suggest that two further keys to understanding an underlying design in the biochemistry of life are **porphyrins** together with the root word meaning of the one word – 'electron'. I will use italics and boldface text for significant etymology from here on.

***Very strangely*** (so I detect God's hand at work) ***'electron' has an etymological meaning of the amber of a tree*** – what comes out of a tree – it will be hugely significant.

Porphyrins are organic compounds used widely in biochemistry and are composed of four pyrrole rings connected by methine bridges. Figure 3.1 shows how haemoglobin and chlorophyll possess porphyrins within their structure (there are other components to them – a phytol chain in chlorophyll, and polypeptide chains for haemoglobin).

***Porphyrin has a root word meaning of 'purple'.***

Purple in the Bible is associated with royalty, high status and, most significantly, the Lord Jesus Christ prior to His crucifixion, where soldiers put a purple cloak on Him to mock Him. Purple was also the colour chosen in the temple to symbolise God or unite God and Man together. I suggest pigmented porphyrins are representing the work of the purple-robed Lord Jesus Christ.

**Figure 3.1: Iron heme and chlorophyll; Peter Hermes Furian; Shutterstock.com**

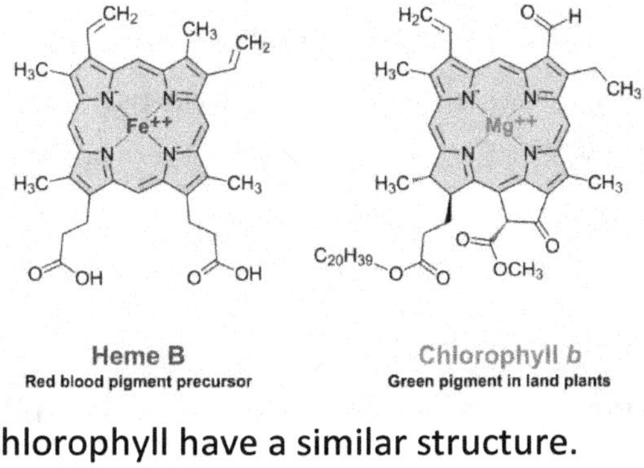


Haemoglobin and chlorophyll have a similar structure.

The structure, function and naming of the porphyrin-based proteins and pigments will reveal much Christ-centred design. The pyrrole rings are the five-sided structure with an 'N' (nitrogen) at one vertex. The other vertices are all carbons (C), which are omitted here to avoid clutter (these five-vertex rings are also termed 'heterocyclic rings'). The nitrogenous bases in DNA and RNA all have heterocyclic rings too, but some are six-sided. I will show, amazingly, how the rings seem to signify a 'whole' living entity and their chemical formula again signifies their contribution to Man's born-again spiritual status, or what they are working on in Man's spirit to achieve his saved status. God's own representation appears to be five-membered rings, and Man and the Bible seem to be six-membered. Please note the diagrammatic similarity here with a 'phosphate group'; I have previously shown how the phosphate group represents a 'light bringer' – the phosphorous lies in the middle – bringing forth four 'oxygens', and I have shown how these represent 'lights'. I believe this applies to porphyrins too. (In reality, please note that the 'heme' and chlorophyll are planar but the phosphate group is not.) Just like the phosphate group, is the centre a representation of a Christ-like trait? (Christ in the middle of a cross.)

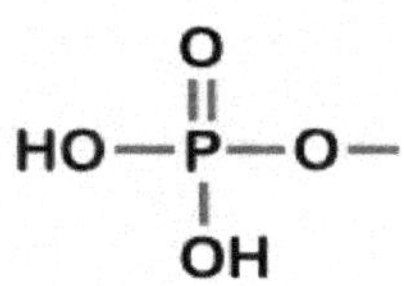


From figure 3.1

***Pyrrole has a root word meaning of 'reddish' or 'fiery'.***

The whole 'square' of four pyrroles joined together with iron in the centre is called a 'heme' group in haemoglobin.

***Heme has a root word meaning of 'blood'.***

Blood is recognised throughout the Bible as a reference to life, covenant, sacrifice and redemption, and as being a major part of rituals of atonement and redemption; especially Christ's sacrifice.

The centre of the porphyrin in haemoglobin is iron (Fe). The centre of the porphyrin in chlorophyll is magnesium (Mg). Haemoglobin makes blood red, and chlorophyll makes plants green (please note the way Christ is likened to a *'green herb'* in Genesis 1:30[23] within the biblical KRM).

***Iron also has a root word meaning of 'metal', 'powerful' and 'holy',*** and is also associated with 'blood'. Iron in the Bible is linked with strength (Christ rules with an *'iron sceptre'* (Revelation 2:27)).

***Magnesium has a root word meaning of a mineral ('lodestone') and also 'with a magnet', (lodestone has a meaning of 'way' or 'guide stone')*** and Jesus Christ is the *'cornerstone'* of the church (Ephesians 2:20[24]); and Jesus Christ, via the Holy Spirit, also 'draws' people in.

I suggest that porphyrins represent Christ's work; and specifically, haemoglobin represents His blood sacrifice and what aspects of salvation are achieved by this sacrifice. Chlorophyll represents Christ building His church (gaining converts) and what aspects of salvation contribute to this building process.

I will show how the five-sided heterocyclic rings in porphyrins, with four carbons and one nitrogen, are significant; and I will show that each porphyrin having four of these rings (or groups) is also significant. On top of this, haemoglobin has four (heme) groups, which is also important; and, amazingly, much usage of these compounds in biochemistry will be shown to adhere to the text of Genesis 1 and 2. I will consider other porphyrin-based proteins too – notably pheophytin, cytochromes and myoglobin (myoglobin is similar to haemoglobin but does not have the quaternary or four-heme structure).

Please note that the four pyrrole rings are joined by a methylene bridge ($CH_2$).

---

[23] Genesis 1:30 (KJV): ***And to every beast of the earth, and to every fowl of the air, and to every thing that creepeth upon the earth, wherein there is life, I have given every green herb for meat: and it was so.***

[24] Ephesians 2:20 (KJV): ***And are built upon the foundation of the apostles and prophets, Jesus Christ himself being the chief cornerstone.***

***Methylene has a root word meaning of 'wine' and 'wood'*** (methy = 'wine' and 'hyle' = 'wood'). In the Bible, Christ's blood is often symbolised by wine and His crucifixion as being on wood – the pyrrole rings are joined by Christ's blood sacrifice, perhaps.

Please always bear in mind throughout this book the absolute fundamental that:

**'Salvation was purposed by the Father, accomplished by the son and applied by the Holy Spirit'**[25]

To recap: all left-handed or 'L' amino acids and other 'L' compounds and their subsequent peptides/proteins will represent being **made by** the Lord Jesus, or being Christ-like or even a part of *'the body of Christ'* and a *'bride'* of Christ. All right-handed or 'D' sugars/compounds will represent the **working** Holy Spirit. The chemical formula of 'D' sugars and other compounds will reveal what the Holy Spirit is **working on or with**, and in other non-chiral compounds, the chemical formulae will often reveal a specific role or even Man himself – **their etymological 'background' will reveal what role is being referenced, and their chemical formulae will reveal what they are contributing to, or need, to 'manifest' this role.**

[25] A quote from the Gospel Coalition: https://www.thegospelcoalition.org/essay/the-holy-spirit-agent-of-salvation/

# Chapter 4

# Adenosine Triphosphate

*'Ereb to boqer'*

*The energy currency of all life and the story of salvation*

## The Theory

We have already mentioned adenosine triphosphate (ATP) in Chapter 2, but now for a bit more detail. Within DNA, adenine was interpreted to be representing the Lord Jesus Christ. This is confirmed by its chemical formula revealing all the **contributions** that Christ made to

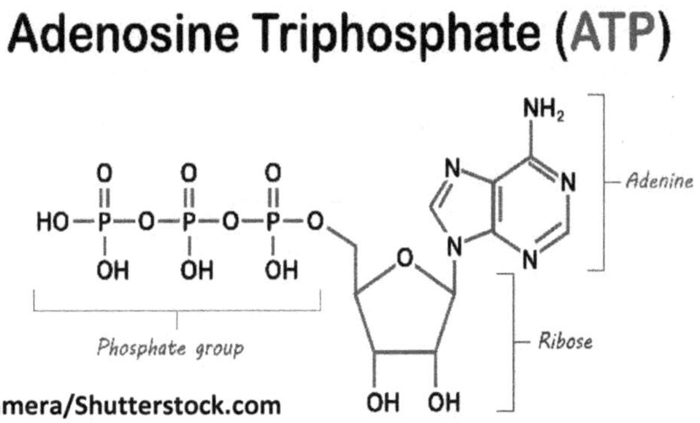


**Figure 4.1: ATP; BigBearCamera/Shutterstock.com**

Man's born-again spirit (confirmed by Genesis in that strange 'elemental-chemical-formula-count way'). Similarly, ribose was interpreted as representing the **working** Holy Spirit, and this is also confirmed by an 'elemental-chemical-formula-count' – what He **worked with** to make Man's born-again spirit.

Thus, within ATP Christ and the Holy Spirit are both represented together by the presence of adenine and ribose (*adenosine* is formed by combining 'adenine' and 'ribose') – **Christ's accomplishment and the Holy Spirit's work**.

The phosphate groups in ATP consist of a 'light bringer' (phosphorous) and four lights (oxygens) and *-ate* has a meaning of 'acted upon'; thus *phosphate* can be considered as 'light bringers acted upon'. The light bringer may be God the Father, Jesus Christ, the Holy Spirit or even saved Man. The lights they bring may be any combination of God the Father, Jesus Christ, the Holy Spirit, the Bible and changed aspects of saved Man's spirit, as all are 'lights'. 'Light bringers' seem to bring themselves forward too, as well as other light sources – all working together, but the 'authority' to work comes from the 'light bringer' or 'P'. Jesus Christ gained this authority by a sacrifice of His life and also passed **some** of this authority on to Man (to be a 'light-bringer helper'). The Holy Spirit may be considered a 'light bringer' or authority with regard to 'bringing' the Word of God (Bible) or Christ to Man.

I suggest that ATP can be considered and 'read' in the same way as demonstrated with DNA and RNA: the working Holy Spirit by Christ's work brings light to Man. But...

**My position is: it is also a profound and biblically aligned '*evening*' (*ereb*) to '*morning*' (*boqer*) statement, and this statement is in the same style as Genesis 1:11-12.** Genesis 1:11-12 have this repeating threefold structure – a set of three categories are brought forth and a set of three categories result[26]. As nucleotides are also based on three compounds (nitrogenous base, ribose and phosphate group), any two nucleotides together will have this same 'two sets of three' design too.

---

[26] Genesis 1:11-12 (KJV): ***And God said, Let the earth bring forth grass*** [1] , ***the herb yielding seed*** [2], ***and the fruit tree yielding fruit*** [3] ***after his kind, whose seed is in itself, upon the earth: and it was so.***

***And the earth brought forth grass*** [1], ***and herb yielding*** [2] ***seed after his kind, and the tree yielding fruit*** [3], ***whose seed was in itself, after his kind: and God saw that it was good.***

Throughout Genesis 1 the days end with *'the evening (ereb) and the morning (boqer) the* n$^{th}$ *day'*. This was found to metaphorically mean 'darkness became light by God's light in Man' in the KRM – a summary of forward progress in the salvation process. I suggest nucleotides in chemical reactions and in bonding do the same – adhering to the Genesis text by also declaring: God's light coming into Man – *'ereb to boqer'* – and at the same time telling the biblical story in the same manner as Genesis 1:11-12 as 'two sets of three'.

In biochemistry, when giving energy, ATP loses one of its phosphate groups – a $PO_4$ – to become adenosine diphosphate (ADP). The bond breakage releases energy to fuel many reactions. Likewise, in the other direction, storing energy, ADP has energy added and becomes ATP.

Please note that **there are no chemical changes in the salvation process. Throughout Genesis, the KRM takes the form of before- and after-salvation descriptions – snapshots of a process – but always going back to the beginning of salvation. Nucleotides do exactly the same, and go back to the beginning and then tell the story to an endpoint...**

If I write out the chemical change from ADP to ATP, we can interpret this chemical change to before- and after-salvation summaries: Jesus Christ (adenine) and the working Holy Spirit (ribose) work with two 'light bringers' and present salvation to Man (the two light bringers are God the Father and the Lord Jesus Christ), and each presents using four lights ($PO_4$). After salvation, Man is now a 'light bringer' himself, so the Holy Spirit is now working with thee light bringers – ATP.

ATP to ADP: the Holy Spirit is carrying three light bringers (God the Father and the Lord Jesus Christ, but also carrying further 'light-bringing' authority from Christ with Him for Man). Man accepts and is given this 'light-bringer' status – the Holy Spirit loses 'one light bringer' and is now just ADP. In both cases darkness becomes light in Man – an *'ereb to boqer'* statement.

Within biochemistry, production of ATP (or similar high-energy molecules) is almost the only reason for the processes I am about to present. Within the KRM, nucleotides become *'ereb to boqer'* or before- and after-salvation statements or summaries, just like Genesis 1 in the Bible.

***The root word meaning of nucleotide is based on 'kernel' or 'seed.'***

To re-emphasis: the Bible is also referred to as *'seed'* (the parable of the sower). This makes nucleotides a bit like the Bible – they both tell a story, just as DNA and RNA told a story – and the etymology of the word 'nucleotide' is again consistent with this representation.

It should also be noted that Genesis 1:11-12 is a concise summary of the whole salvation process; and it is almost as if it is nested within the whole salvation story that runs chronologically from Genesis 1:2 to Genesis 2:1. I will show that nucleotides can sometimes do the same in biochemical processes; appearing in the 'middle' to both play a part in the process but also to tell a simultaneous beginning to end story.

**A Hebrew aside:** strangely, the written Hebrew language does not have vowels but the spoken language does, the vowels are implied by the context and spoken appropriately. This does remind me strongly of the biochemical metaphors we have already considered and are about to consider in more detail in this book; the suffixes *'ate'*, *'ine'*, *'ide'* and *'il'* give meaning but the metaphor often just represents an entity or two entities and it is up to us how we chose a word to connect them.

**A six-aside pitch:** the Star of David has six vertices (two triangles intertwined) and I believe that these symbolise the Trinity God and Man's body, spirit and soul[KK] (one vertex for each). I wonder if this is another reason for the six-sided heterocyclic ring representation (the whole spiritual Man in this case).

But there is another pattern in all nucleotides that is at the heart of this book, and we will come to this later.

# Chapter 5

# Nicotinamide Adenine Dinucleotide

*Man at Day 4*

## The Theory

ATP and DNA/RNA are not the only nucleotide-based compounds we are considering here; other nucleotides tell the same full story of salvation but also emphasise how far along in the process we are and what is now 'active' – using the Christ-described ordering of John 3:3-5 – whether Man is 'born again' (after Day 4 in the KRM), 'born of water' (after Day 5 in the KRM) or 'born of the spirit' (after Day 6).

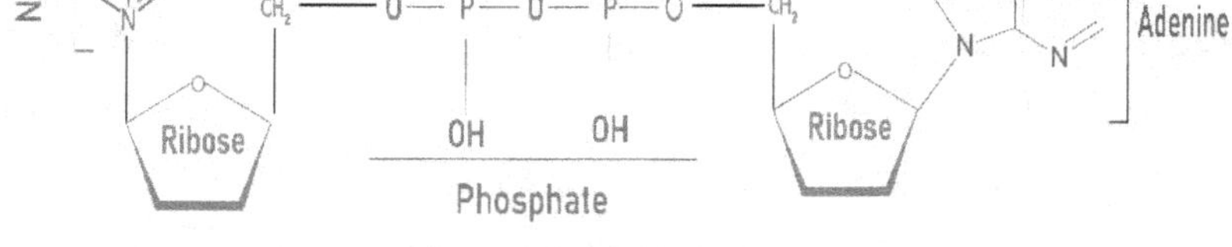


**Figure 5.1: NAD; Sanjay Bartia/Shutterstock.com**

**Nicotinamide adenine dinucleotide** (NAD) is a coenzyme that plays a major role in cell metabolism, especially in 'redox' reactions, where it acts as a carrier of electrons. It consists of two nucleotides 'joined' at their phosphate group end. Nicotine gets its name from Jean Nicot de Villemain, the French ambassador who introduced it into France. 'Nicot' is a derivative of 'Nicholas'.

***'Nicholas' has a meaning of 'victorious people'.***

***'Amide' is derived from ammonia and ammonia is derived from 'ammon' which has a meaning of 'hidden'.*** Colossians 3:3 refers to saved men as being 'hidden' in the *'body of Christ'*. I suggest 'victorious people' with regard to the salvation process are the born again[27]; men after Day 4 in the KRM, saved men functioning in *'the body of Christ'*.

The chemical formula of nicotinamide is $\mathbf{C_6H_6N_2O}$; as with the other nitrogenous bases, can its chemical formula can be logically deduced from the Genesis text? Compare its chemical formula to the pre-saved Man's **contribution** to salvation – thymine ($\mathbf{C_5H_6N_2O_2}$). Saved or 'NAD Man' still has the same inputs into making living waters – $H_6$; and he still has the same input into his likenesses – $N_2$; but he has now gained the image life from Christ that gives him the 'greater light' with its light functionality; the $C_5$ of thymine (the original spirit, the regenerated spirit, the *'beast of the earth'* and the *'body and 'bride'* of Christ) now becomes a $C_6$ because of this continuing addition. This addition is now Man's and not God's. The indwelling of the Holy Spirit in the waters of Day 5 is not exclusively Man's and so is not counted here. The victorious Man was given four lights (the Holy Spirit, the Bible, the greater light and the lesser light) and now they all go together to make this saved Man functioning in *'the body of Christ'* as the *'light of the world'*. This is looking on Man as a 'giver of light' now (just one 'O').

**The chemical formula of NAD is logically deducible from the Genesis text. The guide to its representation is in the etymology of its nitrogenous base's name.** Thus 'reading' the NAD compound from its 'adenine' end: the contributions of Jesus Christ to Man (adenine) with the working Holy Spirit working though other men and their spiritual gifts (ribose) work with four lights (the phosphate group). God the Father is the light bringer, and the lights are God the Father, Jesus, the

---

[27] Romans 8:37 (KJV): ***Nay, in all these things we are more than conquerors through him that loved us.***

Holy Spirit and the Bible. Then, at salvation the four lights are now in Man (the *'greater light'*, the *'lesser light'*, the Bible and the Holy Spirit) and Christ is the 'light bringer'. After salvation *'the body of Christ'* (saved Man is now included) is the 'light bringer' (the 'P' in the phosphate group). The Holy Spirit (ribose) now works with Man's four lights (the *'greater light'*, the *'lesser light'*, the Holy Spirit and the Bible) to make the victorious Man (nicotinamide) a 'light giver' too. Please note how the two phosphate groups share an 'oxygen', allocation of representation to this 'light' will reveal a very significant pattern.

**Chemical formulae from Genesis** (what are the chances?)

When considering DNA, RNA and NAD, I found the chemical formula count for each component to be precise with very little (if any) 'wriggle' room. I even claim that you can predict the chemical formulae.

I like statistics. If I look at things from a biblical perspective then the Bible can mention *'images'* ('C'), *'likenesses'* ('N'), *'living water'* ('H') *and 'lights'* ('O'), and associate then with entities any number of times; The DNA, RNA and NAD chemical element components are in the range 1-10, and I will limit the potential times the Bible 'can' mention them to this number for ease (it is more in reality).

Phosphate ($PO_4$); ribose and deoxyribose ($C_5H_{10}O_4$ and $C_5H_{10}O_5$); adenine ($C_5H_5N_5$); guanine ($C_5H_5N_5O$); thymine ($C_5H_6N_2O_2$); uracil ($C_4H_4N_2O_2$); cytosine ($C_4H_5N_3O$) and nicotinamide ($C_6H_6N_2O$).

That makes for 31 'element counts' being correct in Genesis and each one has a one in ten chance of being randomly so. That gives odds of ten million trillion trillion to one of being entirely random!

{I think that for ATP (and other compounds to come) there is more nuance in the etymology and metaphoric representations – Christ centred, Holy Spirit centred, Man centred, or centred on specific times and events in the salvation process – in these cases the chemical formula can only be explained by Genesis or is 'consistent' with it.}

# Chapter 6

# Flavin Adenine Dinucleotide

*Man on Day 5*

## The Theory

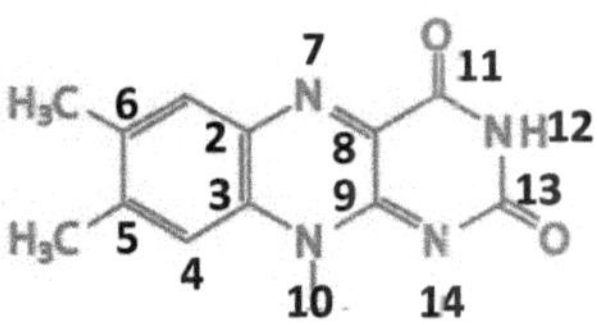


See endnote iii

**Figure 6.1 (right): FAD; Bass Stock/ Shutterstock.com**

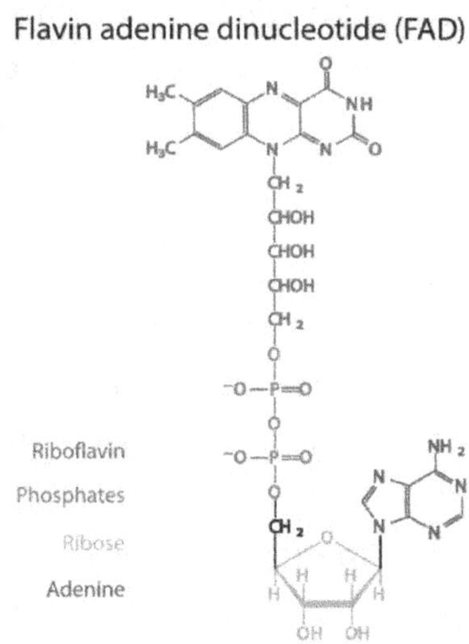


***Flavin has an etymology of 'yellow' or 'golden'.*** 'Golden' is a reference to God in the Bible and notably is used to metaphorise Christ-like attributes in Genesis 2:11-12[28].

Flavin adenine dinucleotide (FAD) is a coenzyme widely used in 'redox' reactions. Please note the way there are now three heterocyclic

---

[28] Genesis 2:11-12 (KJV): ***The name of the first is Pison: that is it which compasseth the whole land of Havilah, where there is gold; And the gold of that land is good: there is bdellium and the onyx stone.***

ring compounds present in the riboflavin 'end', and one consists of just carbon atoms and no nitrogen atoms. Please also note that ribose ($C_5H_{10}O_5$) is replaced with ribitol ($C_5H_{12}O_5$) – I will show in the next chapter in 'additional sugars', how the two extra hydrogen atoms refer to God and Man both working together but counted individually. The 'all carbon' ring is consistent with all Man's God-given image life (see endnote[iii]); and the right hand side ring of the two other six-sided rings matches with the heterocyclic ring (in ring structure element count) of regenerated Man – 'uracil'. The 'uracil ring' contains two likenesses or 'N' elements and I will show that these can represent N2 and N4 as described in Chapter 2. The other 'overlapping' or central ring (representing Man's original spirit being changed) also contains two 'N' elements, and I will show how these can also represent various combinations of 'N' depending on context and perspectives.

Please also note that there are now two lights ($O_2$) on the third heterocyclic ring – this is consistent with two *'the light of life*(s)' being present in Man at this time: both Man's *'light of life'* as described with NAD, and also with an additional light – the light of Christ continuing to pour into Man's spirit to regenerate more and more of it to become more like Christ and ribitol also has $O_5$, in consistency with this. These lights are the summation and not the individual lights now – an end-stage description as with other nitrogenous bases. The whole is consistent with riboflavin representing all God-given-to-Man image life, the Holy Spirit and Man's regenerated spirit all working together – a three-membered 'cooperative'. A possible representation of the FAD ring structure is shown in 'formulae in Genesis endnote[iii].

The three-ring aspect is called an 'isoalloxazine structure' and from left to right the rings are called a 'benzene ring', a 'pyrazine ring' and a 'pyrimidine ring'. The 'ox' comes from oxygen (light or acid-making in the KRM). Uracil is based on uric acid and is looked on as a waste product. We can accumulate all the individual name meanings: 'iso' means the 'same' (note the use of the word *'same'* in Genesis 2:13[29], which we will later consider in depth). 'Alloxan' is based on uric acid and barbiturates (that can also mean something 'foreign'). Benzene also originates as 'something from a tree' and 'pyr' has a meaning of 'fire'. Fire is linked with both the Bible (Jeremiah 20:9[30]) and the Holy Spirit at work purifying saved Man (Acts 2:2-4[31]). Together the whole etymology is consistent with one collective entity containing the three single entities joined together, and these will be working on a fourth spiritual entity, Man's original spirit.

Reading this 'nucleotidal biblical' story from the 'adenine end': Jesus Christ's (adenine) accomplishment and gifts are presented by the Holy Spirit's (ribose) by the authority of a 'light bringer' (Jesus Christ) and presented via four lights. After salvation, the Holy Spirit (now ribitol), Jesus Christ, the Holy Spirit and his regenerated spirit are all working together. Man is now a light bringer too (by the second phosphate group) and has now acquired four specific likenesses (N1-N4). Man's spirit still works with *'good gold'*, *'bdellium'* and *'onyx stone'* twice ($CH_3$ and $CH_3$), again we will cover these three entities in detail.

FAD is also important in its mononucleotide form – flavin mononucleotide (FMN) – and it, too, is a coenzyme and redox component in many metabolic pathways. In the KRM, it can be looked upon as the individuals of Jesus, the Holy Spirit and Man's regenerated spirit; their individual gifting and not from the collective 'cooperative'.

---

[29] Genesis 2:13 (KJV): ***And the name of the second river is Gihon: the <u>same</u> is it that compasseth the whole land of Ethiopia.***

[30] Jeremiah 20:9 (KJV): ***Then I said, I will not make mention of him, nor speak any more in his name. But his word was in mine heart as a burning fire shut up in my bones, and I was weary with forbearing, and I could not stay.***

[31] Acts 2:3 (KJV): ***And there appeared unto them cloven tongues like as of fire, and it sat upon each of them.***

# Chapter 7

# Electrons and Redox Reactions

*The working Holy Spirit*

## The Theory

'**Redox**' is the abbreviation for oxidation/reduction chemical reactions. These reactions both happen at the same time – one thing is oxidised while another is reduced. Oxidation is, as the name suggests, an addition of oxygen, and reduction is the removal of oxygen. But, very counterintuitively, there is a bit more to it: oxidation is also the removal of electrons and the removal of hydrogen (remembered for electrons by 'OILRIG': Oxidation Is Loss – Reduction Is Gain). Much of the energy production in biochemistry involves repeating redox reactions with the same electrons being passed between molecules. In each case, the energy produced from the changing energetic states of the electrons drives energy storage synthesis. The electron movement is, thus, oxidation when it is a loss of electrons and reduction when it is a gain. In cell respiration and photosynthesis, it is an extremely complicated process, with multiple participants in 'electron transport chains' which are sometimes called 'electron transfer chains'.

Within the KRM, there is huge meaning in these reactions. ***To emphasise: 'electron' has a Greek root word meaning of 'amber' or the resin from a tree.*** Amber, or what 'comes from a tree', is an established and repeating metaphor for the Holy Spirit's **work, or more precisely, what comes out of the Holy Spirit** ('*bdellium*' in Genesis

2:12). Thus the Bible itself, spiritual life 'substance' (the Bible terms spiritual substance – *'nephesh chayyah'*) and 'likenesses' (skill sets) may be included within 'electron', as they all are given by or come from the Holy Spirit at salvation (all are covered by 'electron' – tree amber).

***A hydrogen ion ($H^+$) and one electron $e^-$ together make hydrogen (H). 'Ion' means 'going',*** hence '$H^+$' can mean 'water-making going' or 'water-likeness-making going'.

I have described how hydrogen ions ($H^+$) can represent 'units' of 'light' being given. This establishes a necessary link between light and living water, because the products carried by the Holy Spirit (electron) and light 'units' ($H^+$) together make 'H'. But we know 'H' is also used as a parallel for producing living waters – hence the link. This makes for a very elegant parallel to the link between the Holy Spirit bringing forth 'light' and then making 'living waters' flow. Thus, to repeat, for redox reactions: oxidation is oxygen gain, electron loss or hydrogen loss; reduction is oxygen loss, electron gain or hydrogen gain, or in the KRM:

Oxidation is:

- the status of being or becoming a light (O) by a living spirit – an acid maker by etymology
- giving works or gifts through the Holy Spirit (an electron)
- a spirit giving 'units' of light ($H^+$) and then having a status of acquiring a free electron for itself (the Holy Spirit working)
- the giving of 'living water making capabilities' (H) or living water 'likenesses' (also H).

Reduction is:

- the giving of 'becoming a light' status (O)
- the receiving of works or gifts from the Holy Spirit (an electron)
- the receiving of 'units' of light ($H^+$) that when worked on by the Holy Spirit (an electron) will make living water (H).
- the receiving of 'living water-making capabilities' or living water 'likenesses' (also H).

Rather than redox reactions driving energy production, they instead give a very precise picture of the Holy Spirit at work (an electron) giving gifts from God to Man, or using Man to give spiritual gifts, or to form living water. The parallels with Genesis become very apparent and very sophisticatedly presented.

The specific atom that the 'electron' – the Holy Spirit – is associated with will also reveal much; and it is often the oxygen atom that carries electrons or negative charges, representing the Holy Spirit working with 'lights'. Thus, with the use of standard initialisms (please note that a positively charged molecule or area (+) must gain a negatively charged electron (−) to lose that charge and become neutral):

**$NAD^+$ changing to NADH** represents the saved 'victorious' or born-again Man in the '*body of Christ*' gaining something from the Holy Spirit – gifts that contain the 'units' of 'light' that can contribute to living water production. (One electron to remove the positive charge, then another together with a proton (aka $H^+$) to add the hydrogen atom. This applies for the examples below too.)

**$FAD^+$ changing to $FADH_2$** represents Man's regenerated spirit, his indwelling image life and the working Holy Spirit – all working together – Man gaining light to make living waters flow.

**NAD changing to NADP** – the 'P' represents another phosphate group, taking the total to three. This recognises the change in Man's working status or actions to that of a 'light bringer'.

**$NADP^+$ changing to NADPH** represents the saved 'victorious' or born-again light-bringer Man gaining light and Holy Spirit gifts that can contribute to living water production. The 'P' again is symbolising an additional 'light bringer' acting with or being carried by the Holy Spirit.

Actions in the other direction (NADH to $NAD^+$, NADPH to $NADP^+$ and FADH2 to $FAD^+$) just mean Man giving something via the Holy Spirit.

**It is important to note that this is a transaction: the 'victorious Man' or the 'indwellings and Man working together' do not lose their status – they just give some part of it or receive something to add to it.** The 'appearance' of the nucleotide still 'describes' this status; it is the direction of a transaction that is being declared.

**Additional sugars**

I have shown how an element count in the ribose component in DNA matched precisely with the Genesis text and how this changed in the before- and at-salvation ribose – the total element counts stay the same, but for each stage, in the individual parallels, there are slight differences in the makeup of these totals.

If I were to consider the Holy Spirit working before and after salvation (or from both Man's and God's perspective): the working Holy Spirit ($C_5H_{10}O_4$ – deoxyribose) uses four lights (Jesus, God the Father, the Holy Spirit and the Bible – 'the four lights from God') before salvation. The working Holy Spirit at salvation is still $C_5H_{10}O_4$ and also works with four lights in Man (the greater light, the lesser light and the two trees representing the Bible and the Holy Spirit – 'the four lights in Man). Thus, two of these categories apply to both before and at salvation and are repeated (the Bible and the Holy Spirit). If I was summarising the role of the working Holy Spirit both before and after salvation, or from both Man's and God's perspective, it would be $O_6$, as it would cover all six categories.

Jesus and the Holy Spirit provide five contributions to saved Man's 'image lives' as stated in Genesis 1:26-27 (Christ's given image life, the indwelling of Jesus and the Holy Spirit in Man's living waters, and the '*body*' and '*bride of Christ*') – therefore this makes for $C_5$ in 'ribose'. If I consider before salvation conditions: the regenerated aspect of (other) Men's spirit is also used by the Holy Spirit in this process of gaining new converts – of 'multiplying'; but the *'bride'* of Christ is not as this is a personal growth aspect, $C_5$ is maintained but a different $C_5$.

Thus, if I consider the Holy Spirit's role before and after salvation the working Holy Spirit uses the regenerated aspect of Man's spirit, and so the $C_5$ becomes $C_6$ – the working Holy Spirit works with $C_5$ when making Man but $C_6$ if the 'regenerated' aspect of Man's spirit is considered.

The hydrogen element of the working Holy Spirit (ribose) is $C_{10}$, and this corresponded to the rivers in Genesis 2:10-14 (the four river names, three place names and the first place name having three further resources all representing likenesses – making ten in all). The first two rivers (*'Pison'* and *'Gihon'*) are described as having an interaction (by the use of *'kol'* and *'eres'*) – a 'transaction' between Man and God. They can be considered as one transaction or from each individual's viewpoint it becomes two. Although the $C_{10}$ covers the 'transaction' – being made from God to Man in 'ribose' – Man and God can be considered individually when these two 'transactions' are used, for evangelising, hence $C_{12}$ would cover both.

This means that if a named compound (confirmed with D-chirality as representing the working Holy Spirit) was being used to convey the Holy Spirit's work throughout on Man and God (or before and after salvation) its chemical formula would be $C_6H_{12}O_6$. Many will recognise this instantly as the chemical formula of glucose. So, glucose can be used to represent the working Holy Spirit throughout salvation – before and after. Man and God. **This leads to many more parallels.**

**Iron-sulphide complexes** (judgement and purification)

Ferredoxin

**Figure 7.1: Ferredoxin; Malachy666; Shutterstock.com**

In biology, iron-sulphide complexes play roles in enzyme processes and electron transport chains. These are often termed 'iron-sulphur (Fe-S) clusters' – for example, the electron (the Holy Spirit) carrier ferredoxin Fig 7.1.

***'Sulphur' has an etymological history of meaning to burn, or fire, brimstone and hellfire.***

In the Bible, brimstone (generally regarded as sulphur) is associated with judgement and also with purification.[32] We have already discussed how iron is representing the blood of Christ and the amino acids in ferredoxin are representing Christ by chirality. So, logically, I suggest that Fe-S clusters may be representing purification by Christ's blood – an interaction. The compound can be 'read' as meaning that Jesus Christ's works (by chirality of amino acids) present His blood sacrifice (Fe) to sinful Man and purifies him (S).

**Parable of the leaven**

> **Matthew 13:33 (KJV): *The kingdom of heaven is like unto leaven, which a woman took, and hid in three measures of meal, till the whole was leavened.***

I suggest the 'three measures' are very much more significant than realised. I explained in the KRM book how the kingdom of Heaven is more a likeness term – the whole spirit becoming more and more Christ-like in behaviour. I will show how the 'three measures' may represent the indwellings of Jesus, the Holy Spirit and Man's own regenerated spirit (*'good gold'*, *'bdellium'* and *'onyx stone'*) – all working together (leavened) to make saved Man more Christ-like. The '*woman*' is then representing a *'bride'* for Christ. Please note how Zechariah 13:9[33] also uses the term 'a third'.

***'Enzyme' has a root word meaning of 'leaven'.*** Thus, 'enzyme' can reference Christ by chirality (protein enzymes) and when present in a compound name the 'three measures' should be considered.

---

[32] Zechariah 13:9 (KJV): ***And I will bring the third part through the fire, and will refine them as silver is refined, and will try them as gold is tried: they shall call on my name, and I will hear them: I will say, It is my people: and they shall say, The LORD is my God.***

**Conjugate acid-base pairs** (both are the same compound)

Much of the following discussion centres on conjugate acid-base pairs. Conjugate acids are formed when a conjugate base gains hydrogen ions ($H^+$). Conjugate bases are formed when a conjugate acid loses hydrogen ions ($H^+$) – they are both the same compound after losing or gaining hydrogen ions. Hydrogen ions are also known as 'protons'.

The whole metaphoric consideration of these compounds is consistent with 'acidity' ($H^+$) being the currency of 'light' and by extension a component of 'living water'; in each case the change from acid to base is consistent with whatever the compound is metaphorically representing, giving 'light' ($H^+$). This dissociation or deprotonation generally occurs at certain levels of pH, and hence may even suggest that **pH level is representing the presence of light, sometimes at salvation itself – this makes physiological pH, in effect, in some instances equate to salvation. This also makes 'acids' representative of anything that gives 'light'.** The pH in cells will define which version predominates, but you will see it is the conjugate bases that do. (The presence of high acidity (a lot of $H^+$ ions) can be looked on as representing Jesus and the Holy Spirit's own 'light' – conjugate acids give $H^+$ ions in the presence of a lot of $H^+$ ions, in parallel, in the presence of God's light many entities in born-again Man give off 'light'.)

As an example, the conjugate acid, Malic acid **($C_4H_6O_5$)** is deprotonated to its conjugate base – malate **($C_4H_4O_5^{2-}$)** – and both are L-chiral, and hence will represent Christ or the *'body of Christ'* (saved men). This dissociation represents the Christ-like malic acid giving 'light' but still remaining a Christ-like malate. The resulting malate has two 'free' electrons held on oxygen atoms, representing the Holy Spirit working on 'lights'. Please note the name change from *'-ic acid'* to *'-ate'* (a yielder of hydrogen ions to 'something that has been acted on').

Please also note how this process reveals a general image-to-likeness pattern. The malic acid could be considered more of an image statement; but when we have malate with its two working electrons (representing the Holy Spirit working on Man's 'lights') it is more of a likeness statement; so malic acid can be representing Christ-like Man and malate can represent working Christ-like Man – image to likeness.

**I will show how the number of hydrogen ions that deprotonate in different compounds is often predictable from the Genesis text, and it is in the range of 1 to 4 for the basis of the processes covered here.**

**The acid-to-base change and deprotonation in general will always represent forward progress in the salvation process (**please note the phosphate groups of DNA and RNA deprotonate at physiological pH**).**

**A variable key**

Thus, very flexible metaphors are produced. For example, 'living water' may be literal water ($H_2O$) or H. 'Light' maybe be actual light from the Sun, or 'O', or the currency of light – protons ($H^+$ ions). This means 'light' and 'living water' kind of intermingle in this 'micro KRM'.

Any time water is added to a reaction, it may mean Man's living waters are being used; any time water is given off in a reaction, it may mean Man's living waters are being formed.

'C' is image life and 'N' is a likeness, and the working Holy Spirit will reveal what He is **working on** in the chemical formula of what is representing Him by right-handed or 'D' chirality. Other non-chiral compounds may reveal what they **contribute to saved Man** in their chemical formula too, just like all nitrogenous bases in DNA and RNA (and their etymology will reveal who or what they are representing and this may be Man himself).

From a different perspective, the chemical formula can be considered as: what is needed to achieve the functionality of the etymological name; and whilst Joined nucleotides always tell a beginning-to-end story, other compounds may tell an incremental story.

**Generic elements**

Carbon dioxide ($CO_2$) – can represent something that will represent one 'image' and two 'lights' of saved Man's spirit. It could be any 'image life' or any 'light'. It is, in effect, also representing Man himself. $CO_2$ in biochemistry may represent Man's spirit being worked on, and $CO_2$ may represent Man's spirit is added to.

**Generic multi-etymology**

With a compound name like 2-phosphglycerate we have different components for interpretation; I will show how 'gly' will represent Man's saved spirit and we know that 'phospho' means 'light bringer. Given also that 'ate' can mean 'having the quality of' or 'having been acted upon' this gives us several options of interpretation. Are both aspects of '2-phosphglycerate' 'acted upon'? Is it a transaction and the 'light' is acting on the 'saved spirit'? The 'verb' seems to be open to the interpreter. Thus, as 'earth' is a metaphor for Man's spirit in Genesis, then '2-phosphglycerate' on its own can be interpreted to have the same meaning as the biblical 'let there be for lights in the firmament of the heaven to give light ('act on') on the earth (Man's spirit in the KRM)'. The '2' (the specific position of the carbon element that the phosphate group attaches to) can also be an image-lives-used count.

**Generic roles**

**Glyceraldehyde 3-phosphate or PGAL ($C_3H_7O_2PO_4$)** – by its D-chirality will represent the Holy Spirit working with three image lives, seven associations with living water, two lights and one 'light bringer' ($PO_4$). It may be the Holy Spirit who is the light bringer to Man or saved Man himself as a light bringer through the Holy Spirit.

**Amino acids**, and hence proteins, will, by their L-chirality, represent the Lord Jesus Christ's accomplishments, being Christ-like or in *'the body of Christ'*. **Most of the reactions about to be discussed are protein enzyme-driven, and can thus be considered as 'by Christ's work'.**

**What about God the Father?**

The substance of all life can, basically, be divided into four different types of organic building blocks: proteins, sugars (carbohydrates), nucleic acids and lipids. With the parallels of nucleic acids now being shown to represent the Bible – telling a biblical story – we have three of these building blocks 'assigned' to three different divine entities:

- Proteins – the Lord Jesus Christ by L-chirality
- Sugars – the Holy Spirit by D-chirality
- Nucleic acids – the storytellers just like the Bible by etymology
- Lipids (fats) – ?

God the Father repeatedly declares '*the fat is his*' (Leviticus 3:16) or that people should '*make fat*' for Him (1 Samuel 2:29). With this in mind, could 'fat' be representing God the Father? I will not explore this here, but a phospho-lipid membrane may be another way of describing God the Father as a light bringer and indicating that everything enclosed within these membranes is His purpose.

I reemphasise that: **a change from a conjugate acid to a conjugate base will satisfy forward progress in the chronology of salvation.**

***Acid has etymology associated with 'sour', 'sharp', 'rise to a point' or 'pierce' and a 'taste like vinegar'*** (much more on vinegar to come). Please note that the Bible is also described as being as 'sharp as any two-edged sword' and 'piercing' (Hebrews 4:12) and, in a way, the words of the Bible are also a currency of light.

**A non-mistake mistake aside: the 'acid mistake':** Oxygen ('acid maker' in the KRM) is actually a misnomer as it was originally and wrongly thought that it was oxygen that made acidity. However, it fits in perfectly with the biochemical KRM (again, I detect God's hand at work): lights (oxygen) make acid that yield hydrogen ions (water-making capabilities) and leave an electron (the Holy Spirit) often associated with oxygen atoms (with lights).

# Chapter 8

# Photosynthesis Light-dependent Reactions

## *Genesis-day-aligned biochemistry (I)*

**Photosynthesis** means 'making from light'; this is also a nice summary of the salvation process – God remaking Man's spirit from His light. The basic equation for photosynthesis is:

$$CO_2 + H_2O + \text{light} = C_6H_{12}O_6. + O_2$$

In the KRM, this can be interpreted to mean the born-again Man carrying the *'greater'* and *'lesser light'* – $CO_2$ (Man after Day 4), and his living water – $H_2O$ (Man after Day 5), and Christ's light – (light) will make for saved Man working with the 'working Holy Spirit' ($C_6H_{12}O_6$) in the salvation of others, and also Man becoming a 'light giver' himself in the process – shining light ($O_2$).

Respiration may be defined as the production of energy by the intake of oxygen and the production of carbon dioxide. The basic equation for respiration is:

$$C_6H_{12}O_6. + O_2 = CO_2 + H_2O \text{ and energy}$$

(The energy is in the form of ADP becoming ATP.)

In the KRM, this can be interpreted to mean the working Holy Spirit ($C_6H_{12}O_6$) with Christ's 'light' ($O_2$) making born-again Man ($CO_2$); making his 'living water' and making him a light giver (ADP to ATP).

All of this may even be summarised by saying darkness became God's light (in Man) or *'ereb to boqer'* (ADP becoming ATP).

Please note how these equations complement each other – the reactants and products swapping places. The KRM can be applied to both reactions, because glucose can be considered both the end on the salvation process for one Man and the beginning of the salvation process for another. But now it is going to get more interesting, incredibly: I will show the structure of these metabolic pathways matches the Genesis text itself – ***Genesis-day-like biochemistry.***

## The Proof and the Pattern: Photosynthesis Light-dependent Reactions

First, let me state that I am presenting a simplified or perhaps the 'standard' version of some very complicated biochemistry. One of the protein complexes involved in photosynthesis is called 'photosystem II' and it contains pigments, proteins, water splitting clusters and electron carriers; and an average photosystem II complex will contain over 2500 amino acids (**large complexes and enzymes can represent a Christ-driven interaction**). Photosynthesis takes place in chloroplasts – these are organelles within plant cells. These chloroplast organelles have an outer membrane enclosing a thick liquid called 'stroma'. The stroma houses disc-like structures called 'thylakoids', which have their own membranes too. Inside these 'thylakoids' is another liquid substance, this time called 'luman'. These thylakoid discs are stacked like plates one upon another in structures called 'grana'. Photosynthesis light-dependent reactions occur within the thylakoid membranes, and there are location parallels, process parallels and compound parallels.

***'Luman' has a root word meaning of 'light', 'interior space' or 'shining'.*** This can be representing 'light' within the saved Man's spirit.

***'Stroma' has a root word meaning of 'layer', 'covering' or 'substrate'.*** This is consistent with it representing God 'covering Man' – the Bible often refers to Man's sins being removed or 'covered' when Man accepts God's redeeming offer (Psalm 32:1[33]); and Christ's sacrifice covers all the sins that could ever be committed by all men.

***'Manganese' has a root word association with magnetic metal ore and a similar etymology to 'magnesium'.*** I will show how this may be consistent with it representing the Holy Spirit's initial action on Man's spirit – 'drawing Man in' like a 'magnet' (John 6:44[34]).

***'Chloroplast' has a root word meaning of 'making green'.*** This can be interpreted to mean making Man more Christ-like (new Christians and Christ are metaphorised as *'green herbs'* in Genesis 1:30[35]).

***'Grana' comes from 'granum', meaning 'grain' or 'seed'.*** Both of these terms are used in the Bible for the saved or the Bible. 'Grana' will occur in several places in this book.

> **Genesis 1:14-19** (KJV): *Before salvation:* ***14. And God said, Let there be lights in the firmament of the heaven to divide the day from the night; and let them be for signs, and for seasons, and for days, and years:***
>
> ***15. And let them be for lights in the firmament of the heaven to give light upon the earth: and it was so.***
>
> *At salvation:* ***16. And God made two great lights; the greater light to rule the day, and the lesser light to rule the night: he made the stars also.***
> ***17. And God set them in the firmament of the heaven to give light upon the earth,***

---

33 Psalm 32:1 (KJV): ***Blessed is he whose transgression is forgiven, whose sin is covered.***

34 John 6:44 (KJV): ***No man can come to me, except the Father which hath sent me draw him: and I will raise him up at the last day.***

35 Genesis 1:30 (KJV): ***And to every beast of the earth, and to every fowl of the air, and to every thing that creepeth upon the earth, wherein there is life, I have given every green herb for meat: and it was so.*** This text has multiple interpretations for the green herb.

*After salvation:* ***18. And to rule over the day and over the night, and to divide the light from the darkness: and God saw that it was good. 19. And the evening and the morning were the fourth day.***

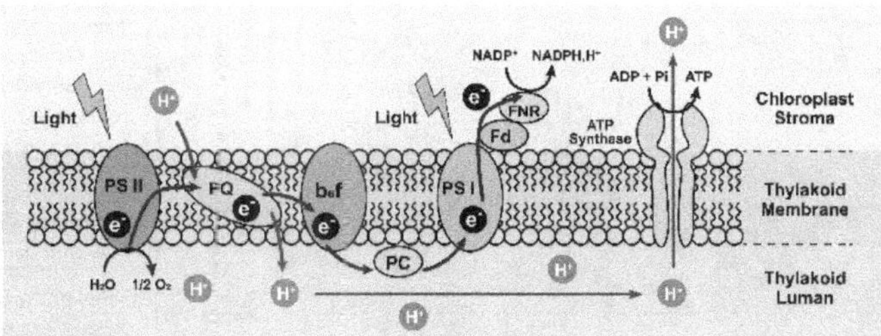


**Fig 8.1: Photosynthesis l ght dependent reactions; Science Photo Library/Alamy Stock Photo**

Basically, photosynthesis involves the use of light energy from the Sun to energise electrons, and these electrons and their energies are passed between complexes or molecules and harvested for energy. In this 'chain of energy release', the energy is used for several roles: to spilt water, to create a hydrogen concentration gradient, and to form high-energy molecules for future energy usage.

In this particular example of the KRM in biochemistry, the 'luman' is the 'light' (or 'goodness' from God) aspect of Man's spirit, the whole 'stroma' is God's role in saving Man (you could say: God in the outside world) and the inter-membrane space is the interaction between Man and God (so, within the inter-membrane space will be representations of both). I will show how photosynthesis adheres very closely to biblical format and chronology. Now in more detail...

### [1] The Word of God

*Before salvation:* ***14. And God said, Let there be lights in the firmament of the heaven***

### [1] Photosystem II

The Sun shines **light** on a large protein complex system present in inter-membrane space in chloroplasts in all plants, called 'photosystem II'. It

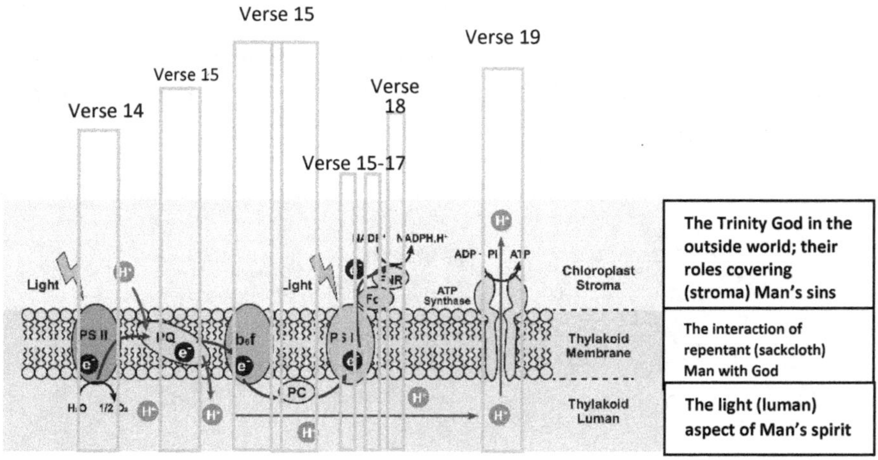


**Fig 8.1; Modified and aligned with Genesis Day 4**

contains many components: **pigment molecules** (carotenoids and chlorophylls) that absorb light; **antennae proteins** (also carotenoids and chlorophylls) that also absorb light and transfer this absorbed light energy to a reaction centre; a **reaction centre** that contains proteins; and a '**special pair**' of chlorophyll molecules (together called 'P680') that absorb the presented energy and become excited by it and start an electron transfer process. There is also an **oxygen evolving complex** on the luman side that contains manganese, and this is involved with the splitting of water to produce oxygen ($O_2$), hydrogen ions ($H^+$) and electrons ($e^-$) – these electrons feed the whole electron exchange process. Photosynthesis begins...

### [1] Photosystem II in the KRM

God the Father is the '**light** bringer' and commands Jesus Christ to be the light of the world shining His light on all men. This light is revealed by four 'lights': God the Father, Jesus Christ, the Holy Spirit and the Bible; **(represented by carotenoids and chlorophylls, and they also represent other saved men)**. The Holy Spirit ('electrons') works with and moves between these lights to present Jesus Christ as the light of the world. The summation of the actions and knowledge of these lights that are to be delivered into Man's spirit will be by two 'lights' – the

Holy Spirit and the Word of God or Bible (**the special pair**). The Holy Spirit begins the salvation process and 'broods' on Man's spiritual waters (splitting) to feed and begin the process (Genesis 1:2, Genesis 2:6 and the **oxygen evolving complex**). The Holy Spirit's action on Man's 'waters' makes Man start to receive 'light' ($O_2$), provides the currency for more light transfer ($H^+$) and gives the Holy Spirit something to work with ($e^-$). Chlorophyll molecules, with their magnesium core, represent Christ being presented to Man as the cornerstone and builder of the church. Salvation begins…

***Pheophytin*** is a porphyrin-based compound just like chlorophyll and haemoglobin, but it does not have a central metal component representing a 'work' of Christ; rather this component is replaced by hydrogen ions ($H^+$) – something is bringing light. ***It has a root word meaning of 'dusky dark plant substance'.*** This is consistent with pheophytin representing the Holy Spirit; and the dusky dark may be the surroundings around sinful men that He occupies in this process of bringing 'light' ($H^+$) – hydrogen ions.

***Plastoquinone: 'plasto' has a meaning of 'to form'. 'Quinone' originates from an 'acid substance from the bark (or 'bark of barks') of the 'quina quina' tree' (the source of quinine)***. We have already discussed the Holy Spirit being metaphorised by *'something from trees'* as seen with ribose, *'bdellium'* in Genesis 2:12 and trees in general in Genesis 1:11-12. But I suggest it should not be overlooked that 'quin' also means, simply, 'five' – and this can be interpreted as to be referencing the five functions of the *'lights'* as described in Genesis 1:14.

### [2] The Word of God

> 14. … ***to divide the day from the night (1); and let them be for signs (2), and for seasons (3), and for days (4), and years (5):***

Please note the similarities with Genesis 8:22:

**Genesis 8:22 (KJV):** ***While the earth remaineth, seedtime and harvest, and cold and heat, and summer and winter, and day and night shall not cease.***

I suggest all terms in both verses are metaphorising aspects of the salvation process in the KRM. All of the terms in Genesis 8:22 will appear in the KRM and the biological KRM.

**[2] Pheophytin and plastoquinone**

The excited special pair pass the excited electron to pheophytin; pheophytin passes the electron to plastoquinone. Energy is released during this process, and it drives a protein pump to send hydrogen ions ($H^+$) from the stroma to the lumen. The electron moves to the $b_6f$ cytochrome complex and further along the electron transport chain.

***[2]* Pheophytin and plastoquinone in the KRM**

The Holy Spirit is the deliverer (pheophytin and electrons) of the light ($H^+$) that enters Man's spirit. The light is from God the Father, Jesus, the Holy Spirit Himself and the Holy Spirit with the Bible. Man gains this 'light'.

***Plastocyanin is a large molecule made mainly of amino acids but with a copper core that gives it an intense blue colour.*** By the L-chirality of its amino acids, it can represent Jesus Christ, and by its blue colour it can (through the general use of blue in the Bible and especially in the tabernacle[36]) represent Jesus being the bridge between Man and God.

---

[36] Numbers 4:5-13 (KJV): ***... and shall spread over it a cloth wholly of blue, and shall put in the staves thereof.***

***And upon the table of shewbread they shall spread a cloth of blue, ...***

***And they shall spread upon them a cloth of scarlet, ... And they shall take a cloth of blue, and cover the candlestick of the light, and his lamps, and his tongs, and his snuffdishes, and all the oil vessels thereof, wherewith they minister unto it:*** (see next page)

**[3] The Word of God**

> ... ***And let them be for lights in the firmament of the heaven to give light upon the earth: and it was so.***

**[3] Cytochrome $b_6f$**

The electrons are passed from plastoquinone to a cytochrome complex called cytochrome $b_6f$. Energy is released during further electron transfer processes within this complex, and it too, drives a protein pump to send hydrogen ions ($H^+$) from the stroma to the lumen (this will be metaphorising God's light into Man). The cytochrome acts to pass the electron to plastocyanin which is present in the lumen side. Cytochrome $b_6f$ contains heme porphyrins.

**[3] Cytochrome $b_6f$ in the KRM**

The salvation process has now (Verse 15) switched to individual men. The cytochrome complexes are consistent with this as these now contain 'heme' groups and not 'chlorophyll-magnesium' groups – as this is now the sacrificial blood of Christ being presented to Man. There are three types of these cytochrome compounds in cytochrome $b_6f$ – these three thus represent Jesus, the Holy Spirit and the Bible. There are also Rieske iron–sulphur protein clusters that could be seen to be symbols of Christ's blood (Fe) starting to overcome Man's sinful nature (S). The change is now representing 'light' changes within Man's spirit and in consistency the plastocyanin is located in the 'luman' to reflect this. More light ($H^+$) enters Man's spirit. Note the way the red heme and the blue plastocyanin together would make purple...the purple-robed Jesus Christ and this colour features in Numbers 4:5-13[37].

---

Numbers 4:5-13 (cont) (KJV)... ***wherewith they minister in the sanctuary, and put them in a cloth of blue, and cover them with a ... And they shall take away the ashes from the altar, and spread a purple cloth thereon:***

### [4] The Word of God

> *At salvation:* ***And God made two great lights; the greater light to rule the day, and the lesser light to rule the night: he made the stars also. 17. And God set them in the firmament of the heaven to give light upon the earth,***

### [4] Photosystem I

Plastocyanin passes on the electron again – now in a lower energy state – this time into photosystem I (the numbering of the photosystems seems the wrong way around, but it just reflects their order of discovery). A second increment of light from the Sun then enters the photosynthesis process. Photosystem I contains: many proteins (including chlorophylls and carotenoids); a reaction centre, another special pair (these are now collectively called P700); and several iron-sulphur clusters. This time, its primary electron receptor is the magnesium-based chlorophyll, and not the hydrogen ion- (or proton)-based pheophytin. Electrons pass through photosystem I to produce now high-energy electrons due to this new influx of light, and this energy is used to synthesise high-energy molecules for energy use. Some electrons pass back to the cytochrome $b_6f$, complex and this is called 'cyclic phosphorylation' – this will prove to be a significant biblically aligned aspect.

### [4] Photosystem I in the KRM

Man accepts the offer of salvation, and a huge input of the light of life from Jesus Christ enters Man's spirit (a second increment of light). Jesus is the light bringer and the 'light of life' floods in. This light and the Holy Spirit work ($e^-$ exchanges) make for darkness-to-light changes (iron-sulphur complexes); washing Man with light ($e^-$ exchanges) with chlorophylls and carotenoids. The Holy Spirit by this process makes '*two great lights*' – Man's regenerated spirit and an input of Christ's image life (**the special pair,** P700, but a different two now than P680 in photosystem II). The Holy Spirit also 'plants' Himself and the Bible in Man ('**he made the stars also**'), and this is paralleled by some electrons going back to the cytochrome $b_6f$ complex again (cyclic photophosphorylation), and the P700 they are associated with will change and then represent them – yet another different two now (the Holy Spirit and the Word of God). The Holy Spirit works further with Man ($e^-$ exchanges).

**On a Wimbledon tennis aside**, please note how the tennis match in 2010 between John Eisner and Nicolas Mahut had a final set score of 70–68 (a complete outlier of a result – more than twice as long as any other Wimbledon set). These are the same basic numbers that are in the chlorophyll special pairs. The number 700 does occur fairly often in the Bible, but it is the individual digits and their biblical associations that intrigue me.

The photosynthesis base pairs are named after the wavelength of light they absorb. In the Bible, '6' is associated with Man (as he was made on Day 6), '7' is often associated with completeness and good (Day 7 was when God declared everything good and complete), and '8' with new beginnings (8 people left the arc and life began, in effect, on Day 8). Then we come to the etymology of the names of the tennis players: as we have already found, 'Nicolas' means 'victorious people' (the born again) and 'Mahut' means 'elephant keeper'. I will show in Part 2 of this book how God chose the biggest animal to represent Christ's indwelling in Man's living waters – the whale – and I wonder if God is doing the same with an elephant. The elephant is representing the image life given by Christ at salvation (in Day 4), but this time in the non-waters part of the spirit – so the biggest land animal is used as the metaphor. 'John' has a meaning of 'God is gracious', and 'Eisner' has a meaning of 'iron worker'; we have already discussed its link with Christ ruling with 'a rod of iron' and iron's association with blood.

Do the names of photosynthesis base pairs amplify the KRM in biochemistry, and does a tennis match also provide the same metaphor of salvation as described in Day 4? (A sporting-ruah metaphor – and a Jesus-centred summary?) God is using born-again men (Nicolas), and the Holy Spirit is carrying an indwelling 'image life' (He is an 'elephant keeper') for Man to win (Eisner won) the gracious (John) blood sacrifice (iron) of Christ. Or perhaps, the gracious God (John) gave a blood sacrifice (Eisner) to yield saved Man (Nicolas), and with Christ in that saved Man (Mahout now meaning Man himself is an 'elephant keeper').

The year 2010 may also have meaning: '20' is associated with waiting for redemption (Jacob's 20-year service to Laban), and '10' with God's guidance (the Ten Commandments).

I often wonder about the mechanism of such a thing; I cannot see God intervening in the game itself (acting on Man's arm to get some more top spin here or whispering *'go for the lob'* there!). How does He make it happen then? To me it can only make sense if these events are 'if' statements; God gives each person a specific set of skills at birth and if John and Nicolas made the most of their abilities: if they put that extra hour of work in when they were 15 'on a cold and rainy night in Stoke'; then they will have the skillsets to eventually produce this 'biblical' result.

Likewise, for me and this book; if I put in the work; if I listened to all of those Chuck Missler DVDs, if I worked hard at biology in preclinical years, if I listened to two modern-day prophets, then I too, will have that skillset required to write this book. (Please note the total number of pages in my two books add up to 700.)

**[5] The Word of God**

> *After salvation:* ***18. And to rule over the day and over the night, and to divide the light from the darkness: and God saw that it was good.***

**[5] Ferredoxin**

The high-energy electrons are passed to ferredoxin – containing an iron-sulphur redox centre and around 100 amino acids. It accepts electrons from photosystem I's iron-sulphur complexes and passes them on to the ferredoxin-NADP reductase enzyme. The ferredoxin-NADP reductase enzyme converts NADP to NADPH – a high-energy molecule used in the next stage of photosynthesis.

**[5] Ferredoxin the KRM**

The born-again indwelt Man is continually worked on, and given Christ's light by the Holy Spirit ($e^-$ exchanges), and this 'imports' light into his spirit. This time the 'dinucleotide or NAD story' is heralding the new existence of the 'victorious man' – born-again Man. The born-again Man is now a light bringer himself (P), and can rule over darkness and light and differentiate good from evil. Note how NAD now has a phosphate group ($PO_4$) – a victorious Man – a 'light bringer' himself.

**[6] The Word of God**

> *After salvation:* ***19. And the evening and the morning were the fourth day.***

**[6] ATP synthesis**

Equalisation of the hydrogen gradient from the luman to the stroma drives an ATP-synthesis 'machine' within the membrane. ATP is a high-energy molecule used in the next stage of photosynthesis. Photosynthesis is about energy production.

**[6] ATP in the KRM**

Please note how the end stage is now Man emitting 'light' ($H^+$). ($H^+$) is leaving Man and going into the outside world (luman to stroma), and this in turn is being proclaimed with ADP becoming ATP. Man is now the light of the world in the '*body of Christ*'. Light ($H^+$) flows from him in consistency. Darkness to coming light; '*ereb to boqer*' (ADP to ATP).

Please note how photosynthesis starts with the breakdown of water and ends with oxygen being released into the atmosphere; salvation starts with the Holy Spirit 'brooding' on Man's spiritual waters (Genesis 1:2) and ends with Man being the *'light of the world'*.

The apparent disjoint in numbering of the two photosystems ('II' then 'I') may also be significant and not a 'mistake': the 'I' referring to Christ, and the 'II' referring to the Holy Spirit and the Bible as the prime participants in each complex.

**Genesis, point by point, parallels biochemistry – compound for compound, action for action, process for process.**

I suggest it is the chronologies that are the most significant aspect here, pointing to very specific design.

I also think that as this symbolism involves huge molecules that represent the 'interaction' itself, then maybe these molecules can be broken down into components. It may be that letter counts, word counts and gematria counts should be checked for corresponding values between compounds and relevant biblical passages – I will leave that for others to consider.

**With consistent logical and scriptural typing, the light-dependent reactions of the photosynthesis process adhere strictly to the Genesis Day 4 text – each verse and each word.**

**But there is another very significant pattern present here**, if we consider photosystem 1, then four lights come from it (two interpretations of P700 twice) just like the four lights that were evident in the DNA metaphor – four 'ambiguous lights' – all open to different metaphoric interpretations...

# Chapter 9

# The Electron Transport Chain

## *Genesis-day-aligned biochemistry (II)*

### Cellular respiration

Cellular respiration is the process by which cells break down glucose to produce high-energy molecules (ATP); the ATP molecules are used to energise many reactions. Respiration takes place mainly in cell organelles called 'mitochondria'. One part of this ATP production process involves another electron transfer mechanism similar to the light-dependent reactions of photosynthesis; electrons again pass through various molecules and complexes (mainly proteins), and the end result is a large number of ATP molecules. The complexes are named from 1 to 4 in Roman numerals. (Please note that these complexes are also huge; complex I is the largest and in Man contains 45 proteins and about 8,000 amino acids.) This whole process is called the 'electron transport chain' (ETC) or sometimes, the 'electron transfer chain' and it bears significant similarity to Genesis Day 5 and specifically the rivers of Eden.

***'Matrix' has a root word meaning of 'womb'.*** Jesus describes, in John 7:38[37], living water flowing out of saved men's bellies (*'koilia'* is the original Greek word for 'bellies', and also means 'womb').

---

[37] John 7:38 (KJV): ***He that believeth on me, as the scripture hath said, out of his belly** (koilia) **shall flow rivers of living water.***

***'Membrane' has a root word meaning of 'covering'.*** The Bible uses the word 'covering' when describing God removing or paying for Man's sins.

***'Mitochondria' has a root word meaning of 'thread' for 'mitos' and 'grain' from 'chondros' – a 'thread grain'.*** The Bible refers to Jesus' sacrifice and redemption as a 'scarlet thread'[38] and also uses 'grain' to symbolise sustenance, salvation and the Word of God.

***Ubiquinone gets a part of its name from the word 'ubiquitous', which means 'everywhere'.*** The Bible describes the Holy Spirit as being ubiquitous. Ubiquinone is also known as 'coenzyme Q'; 'enzyme' has a meaning of 'leaven', and the Bible specifically links 'leaven' with three 'measures' as previously discussed.

***'Cytochrome' has root word meanings of 'container' or 'something enclosed' ('cyto') and 'colour' ('chrome')*** – an 'enclosed colour'.

## The Theory

To understand this chapter please note that Genesis 1: Day 5 and Genesis 2:8-18 are analogous in what they describe. I will cover this in more detail later but please also note Genesis 1:21 is at salvation and Genesis 1:22 is long term after; Genesis 2:10-14 are also long term.

This is going to be the first example of a very often repeated biochemical and physiological adherence to the Genesis text – a specific ordering that appears throughout Genesis 1 and Genesis 2. This ordering will form the heart of this book and I will introduce it here; its use is so widespread that it dumfounds every time I consider it – (as many as ninety times in Genesis 1 and 2), and everywhere I look in biochemistry and in physiology. Please note that the general order:

---

[38] Joshua 2:18 (KJV): *Behold, when we come into the land, thou shalt bind this line of scarlet thread in the window which thou didst let us down by...*

- new Man and something comes out of him – an export
- new Man and something goes into him – an import
- the Holy Spirit and the Bible together giving gifts
- the Holy Spirit alone.

This pattern occurs in many places and each one is a little different.

## **The Proof and the Pattern** (aligned with the rivers of Eden)

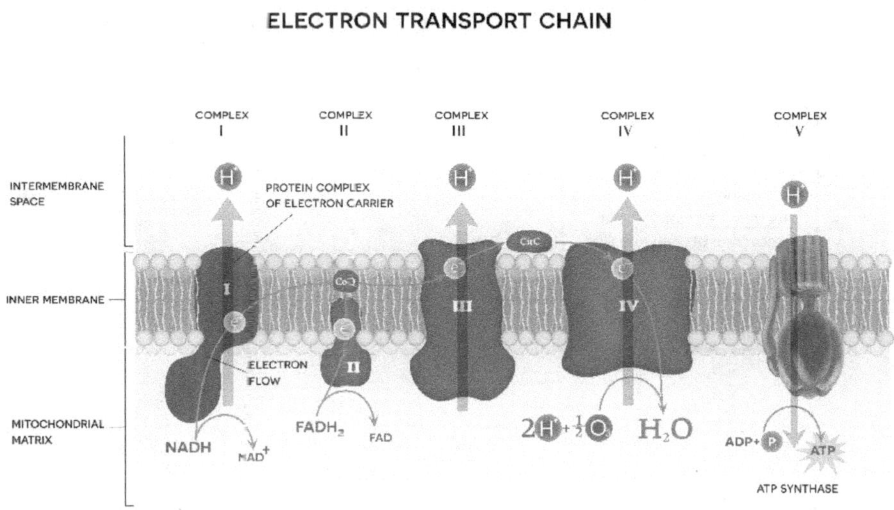


**Figure 9.1: The electron transport chain; TimeLineArtist/Shutterstock.com**

### [1] The Word of God

> **Genesis 2:11-12 (KJV):** ***The name of the first is Pison: that is it which compasseth the whole land of Havilah, where there is gold; And the gold of that land is good: there is bdellium and the onyx stone.***

### [1] Complex I in the ETC

This has the function to oxidise NADH to $NAD^+$ in order to release the electrons needed to start the whole process. The first carrier of the electron (that is removed from NADH) is the mononucleotide FMN. Still within complex 1, the electron passes between various 'prosthetic groups' (non-protein parts in proteins); these include Fe-S clusters. The final electron carrier in the complex is ubiquinone. Energy produced by the electron exchanges fuel a protein ($H^+$) pump; this pumps hydrogen ions from the matrix to the inter-membrane space. This starts to produce a hydrogen gradient that will be used to form ATP.

### [1] Complex I in the KRM (Pison)

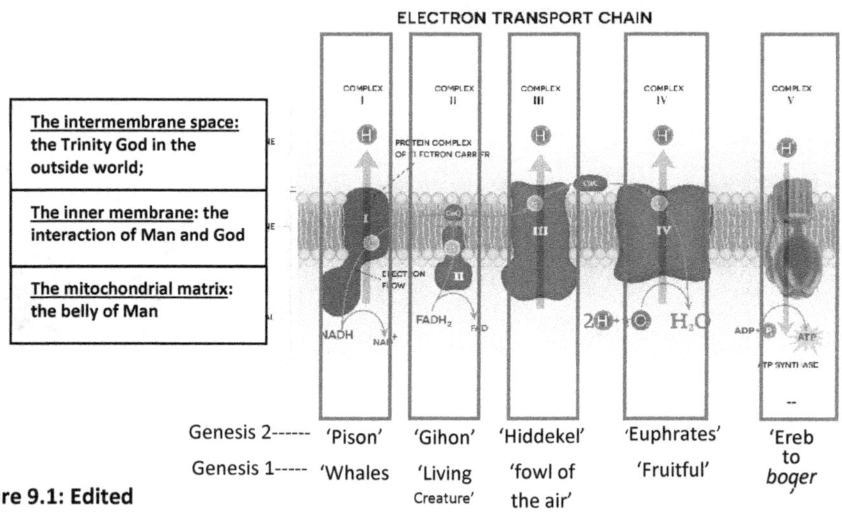


**Figure 9.1: Edited**

The existence of the born-again 'victorious man' (NADH) is recognised at this stage in the salvation process by the appearance of NAD (Man's spirit has already been regenerated at this moment), and the 'victorious Man' is acted on by the Holy Spirit – 'an electron'. The victorious Man gives something (this electron) to the process of making living waters and these in turn 'shine' (give hydrogen ions) into the outside world (**an export**). FMN is the recipient of the Holy Spirit's work (an electron), and this represents the Holy Spirit working on three very specific 'players'. The Holy Spirit moves through Man's spirit, cleaning and purifying, adding and taking away (Fe-S complexes and more electron exchanges). There is an interaction with spirit 'image lives' in the rivers of Eden passage that is referred to as *'compasseth the whole land'* in Genesis 2:11-12[39] (metaphorised by the 'electron'). The resultant 'light' or 'goodness' that this interaction results in, flows out of Man ($H^+$), and this is what goes on to form a part of the living water **(an export)**. The three 'players' are not just represented by FMN, but they are all also metaphorised again in complex I as: Jesus (protein by chirality), the Holy Spirit (electrons and ubiquinone) and Man (NADH, Christ-like Man is represented in amino acids and the sulphur in the Fe-S complexes). The products (**or 'exports'**) of these 'three players' are represented as *'good gold'*, *'bdellium'* and *'onyx stone'* respectively in the rivers of Eden passage (these are all likenesses). *'Pison'* means 'increase' and Man's Christ-like likenesses are increasing.

### [2] The Word of God

> **Genesis 2:13 (KJV)*: And the name of the second river is Gihon: the same is it that compasseth the whole land of Ethiopia.***

### [2] Complex II in the ETC

This has a key role both in this electron transfer chain and also in the Krebs cycle (we will cover Krebs cycle later). Complex II plays the role of oxidising succinate to fumarate in the Krebs cycle. Complex II is given electrons but this time from $FADH_2$, which is oxidised to FAD in the process. Electrons are then passed through various Fe-S clusters before being given to ubiquinone, which moves between complexes I and II but then passes electrons to complex III. There is no proton ($H^+$) pump (export) in complex II (this is significant when considering the direction of flow of light to or from Man's spirit). Complex II also contains FMN.

---

[39] Genesis 2:11-12 (KJV): ***The name of the first is Pison: that is it which compasseth the whole land of Havilah, where there is gold; And the gold of that land is good: there is bdellium and the onyx stone***

### [2] Complex II in the KRM (Gihon)

The existence of the now three-membered 'cooperative' is recognised by the use of FAD at this stage (four-membered if changing Man's original spirit is included). It is acted on by the Holy Spirit (the Holy Spirit acquires its electrons) and then gives something from the 'cooperative' (these electrons) to the process of making living water **(an import)**. Just like other nucleotide use, the $FADH_2$ being oxidised to FAD is a 'biblical type' statement describing a 'light' transaction from the 'cooperative' via the Holy Spirit. The 'cooperative' now interacts with Man's original spirit too (this will be the '*whole land of Ethiopia*') in the rivers of Eden passage; the 'cooperative' consists of the original three members referenced in Genesis 2:12 by the use of 'the same' in Genesis 2:13[40] – Christ's image life and light is now going into Man **(an import)**. The Holy Spirit moves through Man's original spirit, cleaning and purifying, adding and taking away (Fe-S complexes and electron exchanges). The presence of FMN indicates the participants.

Please note that NAD and FAD both feed the same electron pool or 'coenzyme Q', and, in parallel both the regenerated Man and the 'cooperative' in Man contribute to the process of making living waters flow. '*Gihon*' means 'bursting forth' – and this is consistent with referencing abundant gifts, and Man's gifts from Christ are abundant and also bursting forth (in consistency with the use of abundant In Genesis 1:20-21)

### [3] The Word of God

> **Genesis 2:14 (KJV):** ***And the name of the third river is Hiddekel: that is it which goeth toward the east of Assyria.***

---

[40] Genesis 2:13 (KJV): ***And the name of the second river is Gihon: the same*** *(meaning the 'good gold', 'bdellium' and 'onyx' from the previous verse is used again)* ***is it that compasseth the whole land of Ethiopia.***

**[3] Complex III in the ETC**

This receives its electron from ubiquinone and has significant similarities to cytochrome $B_6F$ complex in photosynthesis. It also contains heme-based cytochromes that participate in the electron transfer chain; Fe-S clusters are present and participate too. The final 'mobile' electron carrier that shuttles the electron to complex IV is also a heme-based molecule – cytochrome c. Complex III has a protein ($H^+$) pump that pumps from the matrix to the inter-membrane space (please note the direction in figure 9.1).

**[3] Complex III in the KRM (Hiddekel)**

This has similarities to cytochrome $B_6$ complex in photosynthesis because they are essentially representing the same rationale (and actors) – a presentation by the Lord Jesus Christ's accomplishment (metaphorised by heme porphyrins) making Man's spirit gain light himself – and also emit 'light' ($H^+$) to other men, all carried out by the Holy Spirit. In photosynthesis this process is for salvation itself, and in complex III it is for the giving of gifts of the spirit (more on these later). In both cases the work is by the Holy Spirit and the Bible – hence the similarity – a shared design and **not a shared evolutionary history after all**. The Holy Spirit with the Word of God shares spiritual gifts (electrons) between Man and God: giving gifts from one to the other, purifying Man (Fe-S clusters and the electron transfer chain). '*Hiddekel*' means 'rapid waters' and Man's initial gifts are instant.

**[4] The Word of God**

> **Genesis 2:14 (KJV): *... And the fourth river is Euphrates.***

**[4] Complex IV in the ETC**

This is the final complex in the electron transport chain and reduces oxygen to water. Oxygen, in effect, is the final electron carrier. Complex IV also contains heme-based cytochromes. Electrons are received in complex IV from cytochrome c, and move through it and finally are passed on to oxygen. Complex IV can have a blue colour due to its copper content. This oxygen and hydrogen ions form water ($H_2O$) and free oxygen ($O_2$), which is released into the atmosphere.

The huge protein gradient fuels an ATP-synthesis protein complex – hydrogen ions ($H^+$) flood from the inter-membrane space back into the matrix and go on to form water and oxygen. ADP becomes ATP.

### [4] Complex IV in the KRM (Euphrates)

This represents Man becoming more and more Christ- and Holy Spirit-like ('Euphrates' means 'fruitful') and spiritually growing and evangelising. The Holy Spirit moves through Man and continually presents Christ (hemes) to him and through him to other men (hemes). Christ links Man and God (represented by the blue colour).

Light floods into the belly of Man's spirit, and streams of living water ($H_2O$) then flow out of it. Man shines with the light ($O_2$) of Christ. *'Ereb to boqer'* (ATP to ADP). The Lord Jesus Christ confirms 'streams of living water flow out of Man's bellies' (John 7:38). (Please note the timings and use of FMN and FAD when we discuss the 'three resources of Pison' in the next chapter.)

### Bio-etymological role play

The flow and chronology of Genesis, point by point – parallels biochemistry and physiology – compound for compound, reaction for reaction, and process for process. I suggest again, it is the matching chronologies that are the most significant aspect here, pointing to a very specific design.

**With consistent logical and scriptural typing, the electron transport chain adheres strictly to the text of Genesis Day 5 and the rivers of Eden passage – each verse and each word.**

This strongly suggests the same designer is at work ...

# Chapter 10

# Photosynthesis Non-light-dependent Reactions

## *Genesis-day-aligned biochemistry (III)*

I am now going to describe some very similar but still subtly nuanced variations to the same repeating Genesis framework for the biochemical and physiological KRM. The biochemistry discussed from here on will align with one or more Genesis passages, and the perspective of this alignment will be significant:

- aligned from a *'lights'* perspective (Day 4)
- aligned from a *'made to grow'* perspective (Genesis 2:9)
- aligned from a *'waters'* (and especially the rivers of Eden) perspective (Day 5) and Genesis 2:10-14
- aligned from a *'body and 'bride'* of Christ perspective (Day 6).

## One True Pattern

As well as these passages, I will now re-emphasise the order or chronology that is within all of their verses, an order that has already been mentioned in Chapter 9 and the second half of Chapter 8 (from 'Photosystem I' onwards). The order is: Man's regenerated spirit or what it gives out (an export), Christ's image life in Man's original spirit or what Christ gives to Man (an import), the Holy Spirit with the Bible, and the Holy Spirit alone. Concisely:

**Regenerated Man ∞ Christ in Man ∞ the Holy Spirit+Bible ∞ the Holy Spirit**

This order is hugely significant and will feature throughout the biochemistry of all life; I will go into this in detail later in this chapter, but for now please note that this ordering occurs throughout Genesis 1 and 2 (at least 90 times) but it occurs most notably in: Genesis 1:12, Genesis 1:16, Genesis 1:21, Genesis 1:25, Genesis 1:28, Genesis 1:29, Genesis 2:7-8, Genesis 2:9, Genesis 2:10-14, Genesis 2:15 and Genesis 2:19-20. This all forms part of a very elaborate structure to the KRM in Genesis, which I have named **'one true pattern'**. It is only with understanding of the KRM that the shared biblical metaphoric meanings and shared patterns in the biochemistry of all life can be seen, and this is, I believe, why no one else has recognised it before.

## The Theory

Photosynthesis non-light-dependent reactions are aligned with Genesis Day 5. This parallel takes place in the stroma of chloroplasts.

> **Genesis 2:10-14 (KJV):** ***And a river went out of Eden to water the garden; and from thence it was parted, and became into four heads.***
> ***11. The name of the first is Pison*** (increase)***: <u>that is</u> it which compasseth the whole land of Havilah, where there is gold;***
> ***12. And the gold of that land is good: there is bdellium and the onyx stone.***
> ***13. And the name of the second river is Gihon*** (bursting forth)***: <u>the same</u> is it that compasseth the whole land of Ethiopia.***
> ***14. And the name of the third river is Hiddekel*** (rapid)***: <u>that is</u> it which goeth toward the east of Assyria. And the fourth river is Euphrates*** (fruitful).

Please note how '*Pison*' and '*Hiddekel*' are referred to as '*that is*', but '*Gihon*' is described as '*the same*', and '*Euphrates*' just has the river name alone. There are very precise reasons for this within the KRM; as previously said, I suggest the use of *'the same'* for *'Gihon'* can be referring to the 'good gold', *'bdellium'* and *'onyx stone'* ('the three resources') representing the living entities attached to *'Pison'*; but it has syntactic ambiguity in that it could also be referring just to *'Gihon'* itself (in effect saying: *'Gihon'* contains or carries out...)

These three resources represent what Jesus Christ gives ('good gold'), what the Holy Spirit gives (*'bdellium'*) and what the regenerated aspect of Man's spirit gives (*'onyx stone'*) – three likenesses from three different 'image lives' (this is because Christ is associated with *'gold'*, *'bdellium'* comes from a tree and *'onyx'* with its pink, or 'whitened flesh', is associated with Man – the 'lively stones' of 1 Peter 2:5).

So, *'Gihon'* with its use of *'the same'* can be using these same three image lives and accompanying likenesses, but this time they are acting together as 'one' entity. This again will be very significant in measuring who gets 'credit' for certain actions without 'counting them again' as a likeness in general, and this is also another aspect of a mechanism whereby chemical formulae can be confirmed by the Genesis text. Please also note that *'good gold'*, *'bdellium'* and *'onyx stone'* are **just present** in what is *'compasseth[ed]'* in *'Havilah'* but they actually do the *'compasseth[ing]'*) in *'Gihon'*. This is consistent with the former being just a **'presence'** and the latter being more **'active'**.

*'Hiddekel'* 'going east' can be interpreted to mean going directly towards or of God, because in the KRM, in Eden (Man's whole spirit), the indwelling of God (the garden) is located in the east of it.[41] This ties *'Hiddekel'* with *'fowl of the air'*, as these gifts are also known to be 'directly from God' because they are the group of likenesses termed *'gifts of the spirit'* – 'instant' or 'rapid' gifts that do not initially need nurturing. *'Gifts of the spirit'* are described in 1 Corinthians 12 and

---

[41] Genesis 2:8 (KJV): ***And the LORD God planted a garden eastward in Eden; and there he put the man whom he had formed.***

Romans 12:6-8[42] and are discussed in the KRM book. All are 'likenesses' – briefly, they are wisdom, understanding, counsel, fortitude, knowledge, piety, fear of the Lord, knowledge, faith, healing, miracles, prophecy, discernment, tongues, and interpretation of tongues (abilities or skillsets 'coming out' of Man and originating from the Holy Spirit).

The previously discussed light-dependent reactions (the first stage of photosynthesis) occur in all plants, but there are four variations to the next stage. These are called the 'non-light-dependent reactions' of photosynthesis and each occur in **different types of plants** – these are called 'C3', 'C4', 'CAM' and 'photorespiration'. C3, C4 and CAM are also broad terms used to categorise plants in general, and photorespiration occurs in all of these three plant types to a greater or lesser extent. The 'C' numbers refer to carbon counts in the molecule initially worked on in the process (CAM also has a carbon count of four). All variations again, conform to the text of Day 5 or the analogous Genesis 2:9-18:

- *'Pison'* and the *'whales'* of Day 5
- *'Gihon'* and the *'living creature that moves'* of Day 5
- *'Hiddekel'* and *'fowl of the air'* of Day 5
- *'Euphrates'* and *'be fruitful'* of Genesis 1:22 of Day 5.

The biblical timeline and photosynthesis will be shown to match, in that C4 and CAM will have some '**prior work**' with '$CO_2$' and store it in

---

[42] Romans 12:6-8 (KJV): ***Having then gifts differing according to the grace that is given to us, whether prophecy, let us prophesy according to the proportion of faith***

***Or ministry, let us wait on our ministering: or he that teacheth, on teaching;*** (see next page)

(cont) ***Or he that exhorteth, on exhortation: he that giveth, let him do it with simplicity; he that ruleth, with diligence; he that sheweth mercy, with cheerfulness.***

a compound called malate. This will match with some '**prior text**' in Genesis (Genesis 2:9), whereas C3 photosynthesis will take '$CO_2$' directly from the atmosphere and will match with just the rivers of Eden passage (Genesis 2:10-14) and not Genesis 2:9. Genesis 2:10-14 will also parallel with something called the 'Calvin cycle' (this cycle is for all plant types). The photorespiration aspect will correspond to Day 6 (discussed in Chapter 13). The reactions discussed below take place in the chloroplast *stroma* ('cover'), 'covering' the repentant Man.

Please note what will be a repeating biblical-like metaphor of sweetness for salvation or an 'indwelt-by-Christ Man' – and note that *glucose* has an etymology of 'the sweet one' or 'sweet wine', the same metaphor is used in 2 Corinthians 2:15 and Ephesians 5:2[43] (and wine was used to symbolise Christ's blood at the last supper). These sugars can, simultaneously, also represent the Holy Spirit by L-chirality (thus metaphorically represent, the 'working Holy Spirit' either working for salvation or with the saved). Within photosynthesis the non-light-dependent process is about synthesising sugars (glucose) for energy; in the KRM it is mainly about making living waters flow out of Man – an export. '*Living waters*' make saved Man more pleasing to God, able to spiritually grow in Christ, be alive with the Bible and able to know right from wrong – to make Man a Christ-like and 'sweet-smelling salvation' (please note my carefully chosen order here).

### More bio-etymological role play

Elements and biochemical compound names will, by their etymology, when viewed in a biblical context, reveal **the role** of an entity they represent within a certain aspect of the salvation process (often by extension of an existing biblical metaphor). That entity will be one of,

---

[43] 2 Corinthians 2:15 (KJV): ***for we are unto God a sweet savour of Christ, unto them that are saved, and in them that Perish.***

Ephesians 5:1-2 (KJV): ***And walk in love, as Christ also hath loved us, and hath given himself for us an offering and a sacrifice to God for a sweetsmelling savour.***

or a combination of, God the Father, the Lord Jesus Christ, the Holy Spirit, Man's original spirit, Man's regenerated spirit, the Bible; or any image life or likenesses given by God to Man. I want to emphasise that this is representing a specific **role or status** and not necessarily the entity themselves; for each compound name I ask myself the following questions: **What is the role? Who or what has that role? Does the compound deprotonate? Where in the salvation timeline are we? Do chirality and chemical formula confirm a representation?**

> Genesis 2:9 (KJV): ***And out of the ground made the LORD God to grow every tree that is pleasant to the sight, and good for food; the tree of life also in the midst of the garden, and the tree of knowledge of good and evil.***

**Meaning in the KRM (**one-true-pattern category number in brackets**)**

***And out of the ground*** – Man's spirit – either his original or his regenerated spirit (which is also made from the *'dust of the ground'*).

***made the LORD God to grow*** – this passage is about spiritual growth.

***every ('kol')*** – an interaction between Man and God.

***tree that is pleasant to the sight (***of God***) (1)*** – something **coming out of Man – an export** – Christ-like attributes – associated mainly with the *'lesser light'* or regenerated aspect of Man's spirit.

***and good for food (2)*** – something is good for food and **going into Man himself – an import** – spiritual inputs from Christ – associated with the *'greater light'* or Man's spirit now containing more and more image life from Christ and a continued 'regeneration' of Man's spirit.

***the tree of life also in the midst of the garden (3)*** – a combination of the Holy Spirit and the Bible together; and both are needed to give life. The text can mean that *'kol'* is no longer applying from here on.

***and the tree of knowledge of good and evil (4)*** – the Holy Spirit alone.

Please note the way the trees are made to grow and so can grow at any time in the salvation process – at salvation or long-term afterwards. I wish to re-emphasise how the first two trees ('1' and '2') are covered by the word *'kol'*, and in the KRM book I describe how this is a two-way interaction. The last two trees ('3' and '4') are not. With '1' and '2', the interaction is with Christ via the Holy Spirit, but the last two 'trees' have more 'independence' with 'individual' light-giving functionality; please note that, with *'ground'* included, there are five 'players' described. Please also note the similarities with the rivers of Eden in that the first two rivers, *'Pison'* and *'Gihon'*, are also associated with *'kol'* but *'Hiddekel'* and *'Euphrates'* are not. Additionally, please note that the *'whales'* in Genesis 1:21[44] are also **not** covered by *'kol'*, and it would seem that they should be, because they are equivalent to *'Pison'*. This is because Genesis 1:21 is focussed on created images – Man's regenerated spirit is not a new creation but is a remake, and Genesis 1:21 deals with created images only, and so the regenerated spirit is not included – so there can be no interaction (*'kol'*) with it.

Please note: the participants in Genesis 2:9 (Man's regenerated spirit, Jesus Christ, the Bible and the Holy Spirit) match with the nitrogenous bases in RNA (uracil, adenine, cytosine and guanine respectively), and both relate to after-salvation events. More concisely, the order is exactly as previously described as 'one true pattern'– this can also be looked on as a kind of 'born-again sequence' or '**birthing sequence**'– a very precise ordering – the order of the mechanics of salvation. How can it possibly manifest in biochemistry?

**Regenerated Man (1) ∞ Christ in Man (2) ∞ the Holy Spirit+Bible (3) ∞ the Holy Spirit (4)**

First, some more specific examples of this sequence as they appear in Genesis:

[44] Genesis 1:21 (KJV): ***And God created great whales, and*** ***<u>every</u>*** (kol) ***living creature that moveth...***

- the Genesis Day 4 'lights':
  One great light ∞ another great light ∞ one star ∞ another star
- subject and order of Genesis 2:9:
  Pleasing to the sight ∞ good for food ∞ tree of life ∞ tree of knowledge of good and evil
- Pison ∞ Gihon ∞ Hiddekel ∞ Euphrates

I have already shown and repeat here how this matches with events in the biochemistry of life, for photosynthesis:

- One of base pair 700 ∞ another of base pair 700 ∞ one of the results of photophosphorylation to base pair 700 ∞ another of the results of photophosphorylation to base pair 700

In the electron transport chain:

- Complex I ∞ complex II ∞ complex III ∞ complex IV

And I am about to show how these will also match with:

The compounds in malate synthesis:

Oxaloacetate ∞ malate ∞ pyruvate ∞ phosphoenolpyruvate

The entire Calvin cycle (the initialisms will be explained later):

- RUBP, rubisco, $CO_2$, short-lived compound ∞ 3-PGA and 3 BPG ∞ G3P and glucose ∞ G3P, RUBP and rubisco.

The citric acid cycle:

- citrate and aconitate ∞ isocitrate, oxalosuccinate and α-ketoglutarate ∞ succinate and succinyl CoA ∞ fumarate.

  (And with a 'final' or 'circular' command of: malate, oxaloacetate, pyruvate and acetyl CoA – more to come on this)

Why does God use this pattern? The big statement:

**Jesus Christ: *'I am the resurrection* (1), *and the life'* (2), *he that believeth in me* (3), *though he were dead, yet shall he live* (4);**

(John 11:25)

Because it is the biggest story in the history of mankind; a story that reverberates throughout history and a story that we can now say is present in the biochemistry of all life; the story of how God's loving redemption plan for Man was delivered through Jesus Christ:

Jesus Christ had a glorified bodily resurrection (Luke 24:6[45]) – **(1)**; after that Christ Himself 'breathed' on the disciples **(2)** and told them to accept (in the near future) the Holy Spirit (I believe this is describing Christ's spirit being given to Man by the Holy Spirit, as Christ did the 'breathing', *John 20:22*[46]). Then, the Holy Spirit indwells the disciples at Pentecost **(3)** and they speak in tongues heard in different languages by those around (I believe this represents the Holy Spirit bringing the Bible as that too, is heard in all languages; Acts 2:1-31); and finally, the Holy Spirit will be a teacher and guide to Man (John 14:26[47]) – **(4).** This order is in all of creation to acknowledge the vastness of this moment:

**Regenerated Man (1) ∞ Christ in Man (2) ∞ the Holy Spirit and the Bible (3) ∞ the Holy Spirit (4)**

**In abundance**

For the malate parallel and for other cyclical reactions, it should be noted that cyclical reactions themselves are more efficient than linear ones, as, in linear ones the reactants and products 'get in each other's way'. Hence, all cycles are consistent with the *'abundance'* of Genesis 1:20-21 and this is also consistent with Jesus declaring that Man might *'have it (*life*) in abundance'* in John 10.10. I will show that there is also a similar cyclical nature in the Genesis text itself, 'back to the start'.

---

[45] Luke 24:6 (KJV): ***He is not here, but is risen: remember how he spake unto you when he was yet in Galilee,***

[46] John 20:22 (KJV): ***And when he had said this, he breathed on them, and saith unto them, Receive ye the Holy Ghost.***

[47] John 14:26 (KJV): ***But the Comforter, which is the Holy Ghost, whom the Father will send in my name, he shall teach you all things, and bring all things to your remembrance, whatsoever I have said unto you.***

Genesis 2:8-18 is looking at Man from an image, lights and living waters viewpoint – likenesses are included within 'the living waters' – and not as nitrogen (N) in chemical formulae. 'N' and 'H' are different representations and perspectives for the sometimes-same skill sets, and the parallel chemical formulae used here reflect this living water slant. The nitrogen count appears to be more from a Man's-dominion perspective – his own ownership of certain skills, whereas, the waters are a joint (or precursor) perspective. 'N' replacing 'H' will be a subtle way of defining spiritual growth as Man acquires his 'N's.

In Genesis 2:10-14, God is describing the source of four rivers (an unusual source as they seem to start as full-blown rivers?). Please remember that these are actual rivers that existed at that time and they were named by Man, and also please note that there were certain characteristics to these rivers too, they were not just made up.

***And [1] a river went out of Eden** (Man's original spirit**) to [2] water the garden** (the indwellings and Bible**); and from thence (**after salvation**) it was parted, and [3.4.5 and 6] became into four heads.***

As previously discussed, we can do a count of who contributes to making living waters flow: Man – six; the indwellings of Jesus, the Holy Spirit and the Bible – five; and regenerated Man – four. These get credit for making waters flow, but not all of the 'water likenesses' or gifts that these waters give Man can be credited in the same way. But with knowledge of the KRM, the likenesses can also be individually attributed to each of the 'players' and this will reveal many, many chemical formulae. (From here they will be known as: 'Man's six waters to flow', 'God's five waters to flow'; and 'Man's four waters to flow'.)

**Likenesses**

**The living waters have to flow first before the likenesses 'happen' and this is significant in chemical formulae calculations.** In effect, salvation effects occur after '*from thence*'. The river name's etymology reveal four general characteristics (or likenesses): *'Pison'*; *'Gihon'*; *'Hiddekel'*; and *'Euphrates'*; with specifics attached to some of these:

1. ***The name of the first is Pison*** (an increase)
2. ***that is it which compasseth the whole land of Havilah*** (circle or togetherness), *'compasseth'* declares *'Pison's'* origins or functions.
3. ***where there is gold; And the gold of that land is good*** ('good gold', representing 'from' Christ)
4. ***there is bdellium,*** (representing 'from' the Holy Spirit)
5. ***and the onyx stone*** (representing 'from' regenerated Man)

   'The five water likenesses of Pison' are introduced, numbered one to five, and four image lives are their origin: Man's original spirit (*'whole land'*), Jesus Christ (*'good gold'*), the Holy Spirit (*'bdellium'*) and Man's regenerated spirit (*'onyx stone'*) respectively.
6. ***And the name of the second river is Gihon*** ('bursting forth')
7. ***the same is it that compasseth the whole land of Ethiopia*** (The use of *'the same'* can be a reference to the same three participants as with *'Pison'* (the sources of *'good gold'*, *'bdellium'* and *'onyx'*) but now acting as one, and again interacting (*'whole'* -*'kol'*) with Man's original spirit (*'land'* – *'eres)'*. Jesus and the Holy Spirit are working with regenerated Man and in Man's whole spirit.) *'Compasseth'* can describe how *'Gihon'* 'makes' *'Ethiopia'*.
8. ***And the name of the third river is Hiddekel*** (*'rapid waters'*) 'Rapid' has an association with the Bible as that too is defined as *'quick'*.[48]
9. ***that is it which goeth toward the east of Assyria*** and *'east of Assyria'* is a way to reference God (in this case, chiefly the Holy Spirit). *'That is'* declares *'rapid waters'* move *'east of Assyria'*.
10. ***And the fourth river is Euphrates*** (*'fruitful'* – 'give spiritual help'). Please note how no *'name'* is given and no interactions are described; this is a direct blessing from God the Father (it is

[48] Hebrews 4:12 (KJV): ***For the word of God is quick*** (alive) ***and powerful...***

equivalent to: ***'be fruitful'***... of Genesis 1:22 and 28). This is thus, a broader term; the actioner of all fruitfulness is the Holy Spirit and so *'Euphrates'* can be a reference just to Him; but also Jesus Christ, the Bible and other saved men as all can be fruitful too. I take the position that the latter three are in the long term.

Please note that likenesses may 'come out' of Man (externally pointing), or to point 'inwards'. Please also note that 'the three resources' seem passive (when mentioned with 'Pison', they are just a 'presence') as opposed to the more active ***'Pison', 'Havilah', 'Gihon' and 'Ethiopia'*** ('the four active water likenesses').

**Please note that *'Pison'* and *'Gihon'* are Christ centred but *'Hiddekel'* and *'Euphrates'* are Holy Spirit centred and this will also be very significant (perhaps the most significant thing of all) when considering what particular entity an etymology is representing. Please note that *'Pison'*, *'Gihon'*, *'Hiddekel'*, *'Havilah'*, *'Ethiopia' and 'Assyria'* are transactions but *'good gold'*, 'bdellium' and *'onyx stone'* are individual and I suggest act a bit like 'vehicles'.**

**Who gets the credit for making the likenesses?**

- The regenerated Man is the source of *'onyx stone'* and this is included in *'Havilah'*, and also as part of *'the same'* in making *'Ethiopia'*. This makes for two likenesses, but before these, he still has a general contribution to the four aspects of making living waters flow – 4 and 2 (images and likenesses).

- Man's original spirit is involved in making two likenesses (the two *'whole lands'*, and they have a part in *'Havilah'* and in *'Ethiopia'*), and before that, he has a general contribution to the six aspects of making living waters flow – 6 and 2 (images and likenesses). *'Gihon'* is all about pointing internally – into Man's spirit – Christ's image life coming into Man and Man's spirit being regenerated. (God refers to Christ in man and Israel as 'an apple of His eye').

- The **individual** Holy Spirit's work, covered by *'kol'*, plays a role with *'bdellium'* and also plays His part in the 'cooperative'. But if I consider the Holy Spirit in His role attached to the river *'Hiddekel'* only then that is just two likenesses *'Hiddekel'* and *'Assyria'*, and if I consider his whole Holy Spirit-centred role it also includes *'Euphrates'*. Before that, He still has a general contribution to the five aspects of making living waters flow – 5 and 2 or 3 (images and likenesses). Please note the **working** Holy Spirit works with all 'the ten water likenesses' if all of salvation is considered.
- The Bible is involved with *'Hiddekel'* ('rapid waters') in that it is *'quick'*; and *'goeth toward the east of Assyria'* (*'Hiddekel' 'goeth'* to *'Assyria'* and not the other way round). I will show how the Bible as represented by pyruvate is an after-salvation and long-term role only, the Bible has a general contribution to the five aspects of making living waters flow but after salvation only 'Man's four waters flow' applies. As it is in the long term, it can also be considered as part of the 'long-term aspect' of the broad term 'fruitfulness' (*'Euphrates'*), and this makes for three likenesses in total. (The Bible is not named directly in the Christ-centred-categories *'Pison'* and *'Gihon'* and their water likenesses.) This is unlike the credit the Bible gets in the 'formula' for cytosine – applying to all of salvation). Here the Bible gets credit for 4 and 3 (water-flow) images and (water) likenesses; but with cytosine it gets 5 and 3 (water-flow) images and (man's own) likenesses ('N's).

**Quite remarkably, these likenesses will be represented in the hydrogen count change from conjugate acids to conjugate bases – and this will confirm what the conjugates represent in the KRM.** The conjugate acid reveals a contribution to making living waters flow (sometimes aided by certain pivotal likenesses), and the conjugate base will reveal the 'working' entity with its likenesses in action – both are 'H' counts. The acid is image focussed and the base is likeness focussed – acid to base – forward progress.

## The Proof and the Pattern: Malate Synthesis in C4 and CAM Plants *('made to grow')*

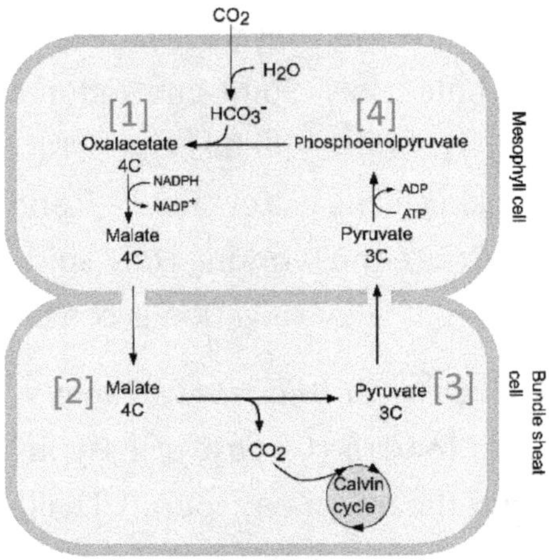


**Figure 10.1: Malate synthesis; Artemide/Shutterstock.com**

***All protein enzymes are words ending '-ase',*** (*-ase* means 'setting apart') and will all represent the work of the Lord Jesus Christ by L-chirality. All of the reactions in figure 10.1 (and most of the ones described in this book) are driven by enzymes.

***Carbon dioxide represents any 'image life' and two 'lights'* – it is achiral**: it may be representing Man's regenerated spirit and his *'greater'* and *'lesser light'*; or it may be his whole spirit with the 'the Holy Spirit carrying the Bible', and the 'Holy Spirit' as lights; or maybe even his whole spirit with its *'light of life'* and more of Christ's light.

*'name'-**ic acid: ic:** 'pertaining to' 'name'; and **acid:*** a 'sharp taste like vinegar' that yields hydrogen ions (light) and leaves a negative charge (the Holy Spirit) in itself.

***-ate***: 'made into', 'processed', 'acted upon'. 'having the quality of'

**Conjugates:** For malate synthesis participants, the conjugate acid will reveal the making of living waters flow (a mainly image statement) and a count of such contributions; and it may also represent one aspect of a 'multiple'-metaphors-etymological name. The conjugate base will reveal a 'dynamic'; and a count of what is used (a mainly likeness statement) in order to achieve the status that the etymological meaning requires, and it too, may also represent a different aspect of a 'multiple'-metaphors-etymological name.

[1] ***Oxaloacetate has an etymology of acid (oxalic acid) and acetic acid (vinegar)* – it is achiral.** Vinegar in the Bible has strong links with Jesus' suffering: He first declined it on the cross, and then accepted it when the suffering threshold was met just prior to voluntarily releasing His spirit. Oxalic acid has an etymology from 'oxalis' or 'acid from wood' – and wood is often a reference to Christ's crucifixion in the Bible (1 Peter 2:24[49]). Oxaloacetic acid and oxaloacetate can, therefore, be considered to be representing Man's spirit giving out light (acid), this in turn is given because of Christ's suffering and death (vinegar). Hence, in a biblical context on Day 5, oxaloacetic acid can be interpreted to be **representing something (the regenerated spirit of Man), which is now a giver of light (a metaphoric 'acid') – and oxaloacetate specifically represents the '*pleasant-to-the-sight*' changes it brings – 'acid from the vinegar'.**

**Oxaloacetic acid ($C_4H_4O_5$) and oxaloacetate ($C_4H_2O_5^{2-}$)**

**[1]** Oxaloacetic acid gives **[2]** hydrogen ions at **[3]** physiological pH to become **[4]** oxaloacetate.

**[1]** Christ-like Man **[2]** gives 'light' at **[3]** salvation **[4]** and this results in 'worked on' Man having 'pleasant-to-the-sight' Christ-like changes.

The conjugate acid, oxaloacetic acid, has $H_4$ in its formula – this is consistent with a representation of the regenerated aspect of Man's spirit making four living waters flow (by the power of the 'acetate').

---

49 1 Peter 2:24 (KJV): ***Who his own self bare our sins in his own body on the tree*** *('xylon' also translated as 'wood')*, ***that we, being dead to sins, should live unto righteousness: by whose stripes ye were healed.***

There are two likenesses coming out of regenerated Man that are 'pleasing to God's sight' (*'Pison'* and *'Havilah'*) – oxaloacetate has $H_2$.

The Holy Spirit continues to work with five lights in the regenerated aspect of Man's spirit (four existing lights and more of Christ's light–$O_5$), but specifically, by the loss of two protons ($H^+$ ions – or giving light) there is a '−2' charge attached to oxaloacetate, with each electron hovering between two oxygen atoms (chemical formulae from Genesis endnote[iv]). This is consistent with the Holy Spirit (each electron) working with two lights (oxygen atoms); the two lights include Man's regenerated spirit (*'lesser light'*) and Christ's light (*'greater light'*) and these are both needed in order to achieve Christ-like likenesses *'pleasing to the sight'* of God.

[2] ***'Malate' has etymology of 'apple'.*** In the Bible, God refers to Christ, Israel and *'the body of Christ'* as the *'apple of his eye'* (Proverbs 7:3[50] and Psalm 17:8[51]) – a term of loving endearment. Man becomes an *'apple of the eye'* when he is in the *'body of Christ'*, and hence just like Christ. Thus, 'malate' represents Christ's image life in Man, and Man in the *'body of Christ'* – this is Man's whole spirit being worked on, given aspects of Christ that make him the *'apple of God's eye'*.

**Malic acid ($C_4H_6O_5$) and malate ($C_4H_4O_5^{2-}$) – Christ in Man**

**[1]** Malic acid gives **[2]** hydrogen ions at **[3]** physiological pH to become **[4]** malate.

**[1]** Man is given Christ-like image life **[2]** and can give 'light' **[3]** at salvation **[4]** to be an *'apple of the eye'* in *'the body of Christ'*.

Malic acid's chemical formula is **($C_4H_6O_5$)** and malate's is **($C_4H_4O_5^{2-}$)** and; they are **both L-chiral** and hence are in the *'body of Christ'*.

---

50 Proverbs 7:3 (KJV): ***Keep my commandments, and live; and my law as the apple of thine eye.***

51 Psalm 17:8 (KJV): ***Keep me as the apple of the eye, hide me under the shadow of thy wings.***

Malic acid has $H_6$ in its formula – this is consistent with Man's spirit making six interactions in making living waters flow. There are four Christ-like attributes that make Man an 'apple of God's eye' ('*Pison*', '*Havilah*', '*Gihon*' and in '*Ethiopia*' – or 'the four active water likenesses'. This is confirmed by malate having $H_4$ in its formula.

The Holy Spirit continues to work with five lights in the regenerated aspect of Man's spirit – $O_5$ (as found with oxaloacetate), but specifically, by the loss of two protons ($H^+$ ions – giving light) there is a '$-2$' charge attached malate, with each electron hovering between two oxygen atoms (chemical formulae from Genesis; endnote[v]). This is consistent with the Holy Spirit (electron) working with two sets of two lights (oxygen atoms); the two lights include Man's regenerated spirit and Christ's light and these are both needed in order to achieve Christ-like attributes that make Man 'an apple of God's eye'.

[3] ***Pyruvate has etymology of 'fire' ('pyre') and 'grape' ('uva').*** In the Bible, 'fire' is a reference to both God[52] (and in particular the Holy Spirit) and the Word of God (Jeremiah 23:29[53]); 'grape' is a reference to 'God's provision and blessings'. This is consistent with pyruvic acid and pyruvate representing the Bible in a joint role with the Holy Spirit working on Man **after salvation** (the Holy Spirit alone is also a presenter of 'fire' and the 'grape').

**Pyruvic acid is ($C_4H_4O_3$) – pyruvate ($C_3H_3O_3^-$) – is achiral**

**[1]** Pyruvic acid gives **[2]** hydrogen ions at **[3]** physiological pH to become **[4]** pyruvate.

**[1]** The Holy Spirit and the Word of God give **[2]** 'light' at **[3]** salvation **[4]** to baptise with fire and give spiritual sustenance.

**Pyruvic acid ($C_3H_4O_3$) and pyruvate ($C_3H_3O_3^-$):** Pyruvic acid has $H_4$ in its formula – this is consistent with it making four interactions in

---

52 Hebrew 12:39 (KJV): ***For our God is a consuming fire.***

53 Jeremiah 23:29 (KJV): ***Is not my word like as a fire? saith the LORD; and like a hammer that breaketh the rock in pieces?***

making living waters flow when the Bible is considered after salvation. Pyruvic acid and pyruvate bring the purifying fire ('pyre') and sustenance (grape or 'uva') for Man's spirit. Please note how fire is associated with the Holy Spirit in the Acts 2:4 passage: ***'And there appeared unto them cloven tongues like of as fire and it sat upon each of them**.'*

The Holy Spirit and Bible together 'in this role as 'pyruvate' make for three likenesses: '*Hiddekel*', '*Assyria*' and '*Euphrates*' (this is with the broad term '*Euphrates*' looked on as including the long-term command (N7 from Chapter 2)), and so pyruvate has $H_3$. (Long-term commands appear after 'one true patterns' 11 times in Genesis 1 and 2.)

The Holy Spirit continues to work with the Bible, 'both' living in Man's spirit, and this is consistent with pyruvate's $^{-1}$ charge hovering between two oxygen atoms; endnote[vi].

Please be aware that all of pyruvate's central roles in metabolism are consistent with it acting as a metaphor for the Bible and the Holy Spirit together: **(1)** a glycolysis end product; **(2)** a conversion to acetyl CoA in the Krebs cycle; **(3)** fermentation to lactate; and **(4)** amino acid synthesis. These will represent the Bible: **(1)** as an instrument of light; **(2)** as an initiator of salvation; **(3)** as nurturing Christians; and **(4)** as making Man more Christ-like. These will be discussed in Chapter 14.

[4] ***'Phosphoenolpyruvate' (PEP): by pyruvate, also has etymology of 'fire' ('pyre') and the 'grape' ('uva'), but also incorporates phosphorous – a light bringer; and an alcohol group (enol), with 'alcohol' having etymology linked with a purified substance.*** This is consistent with 'phosphoenolpyruvate' being a reference to the contribution of the Holy Spirit presenting 'purifying fire' and 'sustenance'; but also other purifications of Man; and to bring light to Man and aid Man to bring light to others – the fruitful Holy Spirit. (Please note there is no mention of 'gluing' in the etymology.)

**Phosphoenolpyruvic acid is ($C_3H_5O_6P$) – phosphoenolpyruvate ($C_3H_2O_6P^{-3}$) – is achiral.** Please note that when considering 'lights' (oxygens) in formulae, I will tend to separate the phosphate group from here on. The phosphate group is the 'light bringer' for salvation, whereas, the other 'oxygens' are lights in the spirit of Man, in God or to be given to the spirit of man. The context will define 'lights' or 'light bringers'.

**Phosphoenolpyruvic acid: $C_3H_5O_2(PO_4)$ and phosphoenolpyruvate $C_3H_2O_2(PO_4)^{-3}$.** (Interestingly, phosphoenolpyruvate is one of the highest-energy molecules out there – the 'main-actioner' Holy Spirit'?)

**[1]** Phosphoenolpyruvic acid gives **[2]** hydrogen ions at **[3]** physiological pH to become **[4]** phosphoenolpyruvate.
**[1]** The Holy Spirit gives **[2]** 'light' at **[3]** salvation **[4]** to achieve a role for Man in making living waters, make Man understand the Bible, purify Man, make Man be fruitful in himself and also in others.

Phosphoenolpyruvic acid has $H_5$ in its formula – this is consistent with the Holy Spirit making five interactions for living waters to flow; His 'pyruvic' brings the purifying fire ('pyre'); and brings the sustenance for Man's spirit ('uva'); His 'enol' means working with a purified substance, and 'phospho' represents light bringing. The Holy Spirit is involved with 'five waters to flow' and hence – $H_5$.

The individual Holy Spirit (alone) in the context of the river '*Hiddekel*' only, gives an 'assist' in making two non-'*kol*' and 'Holy Spirit-centred' likenesses ('*Hiddekel*' and '*Assyria*'); this is confirmed by phosphoenolpyruvate having $H_2$ in its formula. Please also note that the Holy Spirit as '*bdellium*' is not included as that is a part of 'Christ-centred' or '*kol*' likenesses.

The Holy Spirit continues to work with lights in Man's spirit, and this is consistent with a '$-3$' charge on phosphoenolpyruvate. The negative charges in phosphoenolpyruvate are located: one on the pyruvate element and two on the phosphate group. This is again consistent with the Holy Spirit working with Man's 'light of life' in the saved Man, but, as described in Chapter 2, also with the four lights used in saving other men (the regenerated Man or God the Father, the Holy Spirit, the Lord Jesus Christ and the Bible again). Formulae from Genesis endnote[vii].

In conclusion: malate synthesis adheres to the structure of the KRM and is very similar to Genesis 2:9, which, in turn, adheres to the structure of the rivers of Eden – birthing sequences all:

**Regenerated Man ∞ Christ in Man ∞ the Bible ∞ the Holy Spirit**

**Oxaloacetate ∞ malate ∞ pyruvate ∞ phosphoenolpyruvate**

As well as this, the concomitant energy production or usage reactions are also consistent with this parallel (please refer back to figure 10.1):

- NADPH gives off light to become NADP – '*pleasing to the sight*'.
- NADP gains light to become NADPH – the '*apple of God's eye*'.
- God and the Bible give light to Man (ATP becomes ADP).

Please note the timing of the nucleotide transaction and how it precedes the change it brings. Before 'light bringing capabilities' are given ATP becomes ADP; and NADP gains or loses an 'H' before the reversible oxaloacetate to malate enzyme-catalysed reaction.

**Mainly sweetness and light**

Please note again that many of the metaphors in this biochemical KRM are plays on being in Christ ('sweet' and 'an apple') and 'light'. Light is the goodness of living entities (metaphorised by phosphorous, oxygen, acids and hydrogen ions). There will, especially, be many variations derived from 'gly' and 'glu', meaning 'sugar'. *-Ose* itself can also mean 'full of' which is again consistent with its metaphoric meaning.

**God is just using the same metaphors in the biochemistry of all life as He does in the Bible; and with specific etymological meaning that ties in with specific roles.**

## **The Calvin Cycle** (a rivers of Eden perspective)

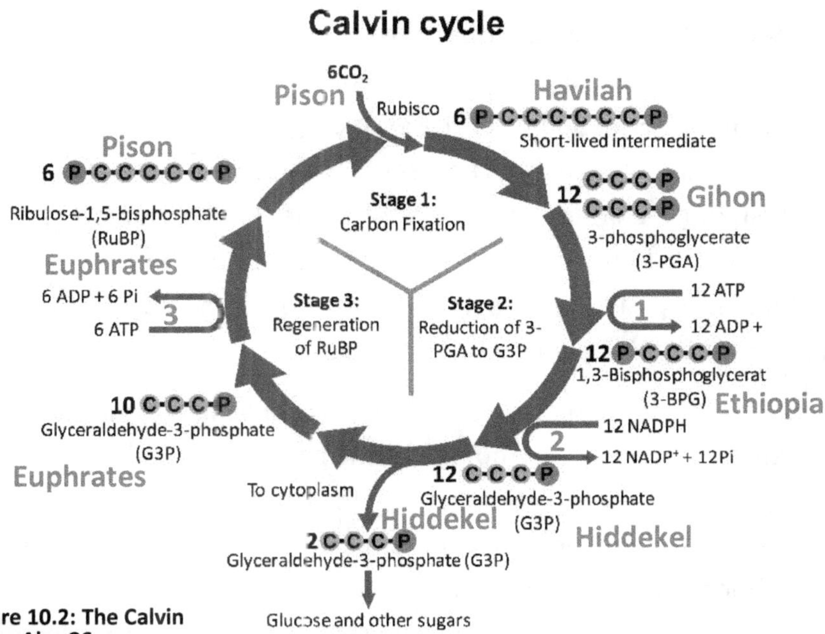


**Figure 10.2: The Calvin cycle; Akor86; Shutterstock.com**

The Calvin cycle is a series of reactions in photosynthesis that fix carbon dioxide gas from the atmosphere and turn it into organic compounds. It also appears to adhere strictly to the Genesis text of 2:10-14 – the rivers of Eden. The Calvin cycle takes place in the stroma of chloroplasts – God covering Man. 'Calvin' has a meaning of 'baldness', and baldness is associated with being clean in the Bible.[54] The Holy Spirit is the vehicle transferring 'light', and the Calvin cycle presents this process from the Holy-Spirit-carrying-this-light-bringing-task viewpoint.

### Consistent 'formulaic' rules

Please note that many compounds considered here have more than one name and I have always focussed on the commonest one. I showed in Chapter 2 how, when considering the nitrogenous bases of DNA, the

---

[54] Leviticus 13:40 (KJV): ***And the man whose hair is fallen off his head, he is bald; yet is he clean.***

'water flowing' count was used but when the working Holy Spirit by chirality (ribose) was considered it was the likenesses that were counted. In that instance, 'ribose' represented the Holy Spirit by both etymology and chirality; but we now have examples where the chirality represents the Holy Spirit but the etymology may represent something else. The L-chirality always represents what Christ gives and the body of Christ, but D-chirality can reveal what the Holy Spirit made and then works with. When a compound represents both the Holy Spirit and another entity (as etymologically revealed), in all cases, changes to these compounds reveal forward progress after salvation.

In all of these examples the hydrogen count follows the same rules: the entity that the etymology principally represents has a 'full' hydrogen or 'waters to flow' count for the whole process, but also sometimes with 'essential' likenesses required 'to live up to this main metaphoric etymological meaning' added. This will apply to conjugate acids, but the conjugate bases will be likenesses centred – likenesses that are needed to achieve a (sometimes secondary) etymological meaning of the compound.

Maybe it is a representation of living waters flowing as image life and in Man's spirit (acid), and when they burst out of Man and operate in the outside world they are likenesses (base).

**Generic etymological naming**

To repeat: It will become apparent that a compound's etymology may, in a way, be considered generic with the same compound having detailed etymological interpretations that can represent different entities or the same entity at different 'times'. In different contexts each compound or their constituent elements may encompass: subtle differently-nuanced etymology; different 'light bringers'; different image lives; different likenesses; different living waters; and different 'lights'.

**A river: running through it**

The rivers of Eden follow 'one true pattern' with '*Pison*', '*Gihon*', '*Hiddekel*' and '*Euphrates*', but in addition, they add more detail on the first two Christ-centred categories; and the Calvin cycle adheres to this same extra-detail format. It is worth considering the exact meanings of these details first:

'*Pison*', (an increase) Christ's, the Holy Spirit's and regenerated Man's **likenesses** come **out of** Man. These likenesses also interact with Man's whole original spirit ('*whole land*') and wash it clean and they are then '*Havilah*' (together).

'*Gihon*' (bursting forth) Christ's, the Holy Spirit's and regenerated Man's **image life** burst forth **into Man** in abundance. These likenesses interact with Man's spirit ('*whole land*') and regenerate it, and the darkness in Man's spirit ('*Ethiopia*') disappears.

I am going to show how ribulose 1,5-bisphosphate, the intermediate compound, 3-phosphoglycerate and 1,3-phosphoglycerate match with '*Pison*', '*Havilah*', '*Gihon*' *and* '*Ethiopia*', respectively.

**All of the actors in the Calvin cycle**

***Ribulose 1,5-bisphosphate or RUBP ('Pison') has the same etymology as previously described for ribose, and is thus also associated with 'arabinose'/'gum arabic' – a gum from the acacia tree*** – this being a repeating metaphor for the Holy Spirit. The *-ul-* has an association with ketones and ketone with vinegar. The two phosphate groups are consistent with representing the Holy Spirit working with two 'light bringers' (maybe Himself and Man, Himself and Christ or Christ and Man in different contexts). As Christ is associated with vinegar this is consistent with RUBP representing the Holy Spirit, Christ and regenerated Man.

**Ribulose ($C_5H_{10}O_5$) is D-chiral** and carries no charge, and **Ribulose 1,5 bisphosphate ($C_5H_{12}O_{11}P_2$) is also D-chiral** and carries negative charges located in the oxygen part of the phosphate group, a '$^{-2}$'. Each negative charge hovers between two oxygen atoms, and this is consistent with ribulose representing the Holy Spirit working with four 'lights' each from two light bringers.

***Rubisco (an acronym)*** is the most common enzyme in the world and fixes carbon dioxide from the air into organic compounds in all plants. By its L-chirality, it can represent Christ's accomplishment or just Christ.

***The intermediate compound ('Havilah')*** disappears very quickly and seems to be a way of representing sin – the cleaning and purification of Man's spirit by removing the problems – sinful portions disappear very quickly. The intermediate compound parallels *'the whole ('kol') land of Havilah',* as this is also a reference to the cleaning of Man's spirit. It breaks down into two molecules of 3-phosphoglycerate **($C_3H_7O_7P$)** and hence its formula is generally considered, in its deprotonated form, to be **($C_5H_{12}O_{16}P_2$)** but this does vary. Its name is 2-carboxy-3-keto-D-arabinitol-1,5-bisphosphate. The name could be 'read' as: two image lives get light from three entities by Christ's sacrifice, via the Holy Spirit working with the 'four lights in Man', using the 'four lights from God' by the authority of two light bringers.

***'3-phosphoglycerate' (3-PGA) is D-chiral and has an etymology of 'phosphorous' – a 'light bringer' and 'glycerine' – 'a sweet thing'.*** 'Sweetness' as previously described is associated with Christians and Christ.[55] The etymological meaning is consistent with this compound representing the Holy Spirit (a 'phospho') working on the regenerated Man – a 'sweet thing'; and the D-chirality can also represent the Holy Spirit working with Christ (a 'phospho') acting as a light bringer (bringing Christ's image life) **to** the (or 'sweet') Man. But it can be interpreted as working **with** the 'light-bringer' 'phospho-Man' himself as we will see with 1,3-bisphosphoglycerate. (Please note that Man is

---

[55] Ephesians 5:2 (KJV): ***And walk in love, as Christ also hath loved us, and hath given himself for us an offering and a sacrifice to God for a sweetsmelling savour.***

made a 'light bringer' in Day 4, and the rivers of Eden are after that.)

**3-phosphoglyceric acid ($C_3H_7O_7P$) – 3-phosphoglycerate ($C_3H_4O_7P^{3-}$) is D-chiral and is also known as 3-PGA –*'Gihon'***

**3-phosphoglyceric acid ($C_3H_7O_3PO_4$) – 3-phosphoglycerate ($C_3H_4O_3PO_4^{3-}$).** The 3-phosphoglycerate carries negative charges located in the oxygen part of the phosphate groups. Two negative charges are associated with the ($PO_4$), each charge hovering between two oxygen atoms and this is consistent with the light-bringing Holy Spirit working with four lights when He is 'light bringing'; one negative charge is on another oxygen (light). The '3' can be a reference to working with the three image lives that make the 'the three resources'.

**[1]** 3-phosphoglyceric acid gives **[2]** hydrogen ions at **[3]** physiological pH to become **[4]** 3-phosphoglycerate**.**

**[1]** The working Holy Spirit works as a light bringer with saved Man, and **[2]** giving more of Christ's image life and light in abundance [**3**] at and after salvation [**4**] to make the 'acted on' Man gain more of Christ's image life and light. Formulae from Genesis endnote[viii].

***'1,3-Bisphosphoglyceric acid' (1,3-BPG) also has an etymology of 'phosphorous' (the light bringer) and 'glycerine' ('sweetness').***

**1,3-bisphosphoglyceric acid ($C_3H_8O_2PO_4PO_4$) – 1,3-bisphosphoglycerate ($C_3H_4O_2PO_4PO_4^{4-}$ is D-chiral – *'Ethiopia'***

The etymology is the same as **3-phosphoglyceric acid and** it now represents the working Holy Spirit with two phosphate groups – two 'light bringers' – Christ's light and the regenerated spirit of saved Man as a light – light from one to the other – regenerating Man's spirit.

**[1]** 1,3-**Bisphosphoglyceric acid** gives **[2]** hydrogen ions at **[3]** physiological pH to become **[4] 1,3-bisphosphoglycerate.**

**[1]** The Holy Spirit working with two light bringers gives **[2]** 'light' at **[3]** salvation [**4]** from Christ to Man's spirit – a regeneration. (The '3' can again represent the three image lives of the three resources but the '1' can now refer to reacting with Man's original spirit.)

The negative charges again, representatively, hover around oxygen atoms. The image (carbon count) can reflect the image life 'used' to make the etymological meaning (in the conjugate acid) and the image life enhanced (in the base). Formulae from Genesis endnote[xii].

***'Glyceraldehyde 3-phosphate' (G3P) $C_3H_7O_2PO_4$ is D-chiral and has an etymology of: 'glycerine' – 'sweetness', aldehyde – 'alcohol without hydrogen', 'alcohol' – 'a refined substance' and 'phosphate' –'a light bringer'.*** There are two perspectives in that saved Man and the Holy Spirit can both match this etymology as both are 'light bringers' with these particular 'etymologically-defined' traits – the compound could thus, represent the whole saved Man by etymology and the Holy Spirit by chirality; or the Holy Spirit alone by chirality and etymology. Man is, or the Holy Spirit can work on or with, 'a sweet thing that gives off light ('$H^{+}$' ions) and this sweet thing is worked on by three of the four lights present in light bringers (hence, '3'-phosphate).

G3P has two pathways in the Calvin cycle and the first is consistent with representing the Holy Spirit in His role purifying Man's original spirit – making purified waters come out of Man, and gifting Man his gifts of the spirit – making Man a 'bride' by use of three of 'the four lights in Man' (not regenerated Man's 'light' as that is nurtured mainly in the Christ-centred categories).

**[1] Glyceraldehyde 3-phosphate** gives **[2]** hydrogen ions at **[3]** physiological pH to become **[4]** G3P with a negative charge.

**[1]** The Holy Spirit, working with the born-again Man, works on **[2]** 'light' at **[3]** salvation [**4]** to be used in making Man's living waters and in purifying his original spirit. Formulae from Genesis endnote[xi].

The second function of G3P is with the etymology pointing at Man and the chirality still to the Holy Spirit. This is now the 'fruitful' Man ('*Euphrates*') and making Man fruitful both in himself and in others. This is consistent with '*Euphrates*' being used with no qualifications.

**The Calvin cycle is all about the Holy Spirit's role** (by chirality) (Please see figure 10.2)

**Pison:** *'Good gold'* (rubisco), *'bdellium'* and *'onyx stone'* (ribulose 1, 5-bisphosphate) give an increase to Man's spirit through the Holy Spirit, and then work on Man's whole spirit ($CO_2$) to purify it (**Havilah**) and remove all sin (the intermediate compound) and produce likenesses in Man pleasing on the eye of God.

**1** and **3**. God's light comes into Man ATP → ADP; *'ereb to boqer'*, then:

**Gihon:** Man's spirit by the working Holy Spirit (3-bisphosphoglycerate) works with the light of Christ to bring more of Christ's image life, his own image life and regenerated Man's image life into Man's spirit in abundance. The Holy Spirit works on Man's spirit with these three entities to regenerate it (**Ethiopia**).

**2.** NADPH → $NADP^+$: Born-again Man gives light to form living waters, then:

**Hiddekel:** The working Holy Spirit (glyceraldehyde-3-phosphate) by both chirality and etymology now works on Man purifying his nature making it 'sweet', and with the Holy Spirit Himself acting as a light bringer. The products of this are represented by the glyceraldehyde-3-phosphate being used to synthesis sugars (notably glucose and ribose – that represent the Holy Spirit's further work in Man).

**Euphrates:** The working Holy Spirit (glyceraldehyde-3-phosphate) by chirality but with Man by etymology now works with Man to make him fruitful in himself and others. Please note how this 'fruitfulness' when applied to Man working with other Men is evangelising, or on himself, is spiritually growing. When considering working with other men then, in a way, the cycle begins again.

**Making living waters count**

The subtlety of the hydrogen counts are apparent:

Ribulose 1,5-bisphosphate and the intermediate compound both have

'$H_{12}$' in consistency with a representation of the Holy Spirit working on both Man and God as discussed in Chapter7.

3-PGA has '$H_7$' in its acid form (Man's six waters to flow and just one on the likenesses has 'happened' in this river at this stage (*'Gihon'*)) – '$H_7$'; and the base has '$H_4$' and this is consistent with the Holy Spirit with the Christ-centred 'four active water likenesses' being 'sweet'.

1.3-BPG has '$H_8$' in its acid form (Man's six waters to flow and both of the likenesses in this river at this stage (*'Gihon'* and *'Ethiopia'*)); and the base still has '$H_4$' representing the Holy Spirit using the Christ-centred 'four active water likenesses' – a 'sweet-smelling light bringer'.

G3P (with etymology and chirality pointing to the Holy Spirit) has '$H_7$' in its acid form (God's five waters to flow and the two likenesses of this current river (*'Hiddekel'* and *'Assyria'*)). In its deprotonated form it has '$H_5$' representing by chirality the Holy Spirit working with everything that purifies Man; this happens in *'Havilah'* and *'Ethiopia'* (as *'bdellium'*) as well as *Hiddekel'*, *'Assyria'* and *'Euphrates'* – '$H_5$'.

G3P (with etymology pointing at Man, and chirality pointing to the Holy Spirit) has '$H_7$' in its acid form (Man's six waters to flow and the one likeness of this river (*'Euphrates'*)). In its deprotonated form it has '$H_5$' representing the Holy Spirit working with all aspects that make Man fruitful (Christ-like). This is 'the four active water likenesses' and *'Euphrates'*. But because *'Euphrates'* can also be applied to a long term description then this can become centred on Man evangelising on other men. In this case the relevant water likenesses change as the internal pointing *'Gihon'* and *'Ethiopia'* no longer apply but *'Pison'*, *'Havilah'*, *'Hiddekel'*, *'Assyria'* and *'Euphrates'* do – '$H_5$' still applies.

The Calvin cycle adheres to the Genesis text and the 'one true pattern':

**Regenerated Man ∞ Christ in Man ∞ the Holy Spirit+Bible ∞ the Holy Spirit**

**RUBP, rubisco, $CO_2$, short-lived compound ∞ 3-PGA and 3 BPG ∞ G3P and glucose ∞ G3P, RUBP and rubisco.**

# Chapter 11

# Glycolysis

## *Genesis-day-aligned biochemistry (IV)*

**Glycolysis:** the breakdown of glucose for energy.

**Glycolysis:** the breakdown of the Holy Spirit's role in salvation.

### **The Theory** (see figure 11.1)

**Glucose ($C_6H_{12}O_6$) is D-chiral** – by D-chirality, glucose is representing the working Holy Spirit from before- to at-salvation as previously described. ***'Glucose' has an etymology of 'sweet substance',*** and to repeat, the Bible represents the Word of God as sweet (Psalm 119:103[56]) and salvation by Christ as a sweet-smelling savour (Ephesians 5:2[57]); a 'sweet thing' – something from God (Christ or the Holy Spirit) will be in man. The Holy Spirit presents Christ to the whole world – God the Father is the light bringer, and He commands 'the four lights from God' to save men.

**Glucose 6-phosphate is D-chiral** – by D-chirality, glucose 6 phosphate is representing the working Holy Spirit working with God and with Man (Chapter 7). **$PO_4$** is now attached to glucose. The Holy Spirit moves to

---

[56] Psalm 119:103 (KJV): ***How sweet are thy words unto my taste! yea, sweeter than honey to my mouth!***

[57] Ephesians 2:5 (KJV): ***And walk in love, as Christ also hath loved us, and hath given himself for us an offering and a sacrifice to God for a sweetsmelling savour.***

presenting salvation (sweetness) by Christ to the individual Man via 'the four lights from God'. Jesus Christ is the authority or 'light bringer'.

**Fructose 6-phosphate is D-chiral** (fructose has the same chemical formula as glucose) – by D-chirality, it is representing the working Holy Spirit. The change to fructose is representative and has a meaning of 'fruit' (Christ Himself in Man); the light of Christ is presented to Man and so the Holy Spirit has been 'fruitful'; and when the Holy Spirit saves Man and makes two great lights, He will again be 'fruitful'.

**Fructose 1, 6-bisphosphate (another $PO_4$ is added) is D-chiral:** the working 'fruitful' Holy Spirit moves forward to 'at salvation', and is now working with two light bringers – this recognises the light of life from Christ is in Man (the greater light) and this regenerates Man's spirit – the regenerated spirit (the lesser light) – two phosphate groups now.

***Dihydroxyacetone phosphate (DHAP) is non-chiral: 'acetone' has an etymology of a product of the 'dry distillation of acetates'***. Acetates are derived from acetic acid (vinegar), and vinegar is associated with Christ on the cross in the Bible. **Di** means two (two participants bringing one 'light bringer' by 'phosphate'), and **hydroxyl** is a reference to water (H) and oxygen (O) – with the usual 'H' and 'O' metaphoric meanings of these elements applying. Please note how this is not D-chiral now but achiral (Man). I suggest that by the use of 'di' this becomes an elegant way of describing the 'battle' of Christ's light and Man's light within saved Man, with Christ's light eventually winning or '*ruling*' over all things good; at the same time describing Man being the light bringer to His own spirit and ruling over darkness (endnote[ix]).

***Glyceraldehyde 3-phosphate or G3P is D chiral: aldehyde is derived from alcohol*** (as previously described**). 'Alcohol'** has an etymology of a 'distilled' or 'refined substance'. This is consistent with alcohol representing Man's regenerated spirit or purified whole spirit. **'Aldehyde'** has an etymology of 'alcohol with hydrogen removed'. This is consistent with it representing Man's spirit and perhaps not always his living waters element (Day 4 is generally the 'spirit' and Day 5 the 'waters'). Altogether, G3P can now be considered slightly differently than in the Calvin cycle, in that it is now applying to the working Holy Spirit's role with the regenerated spirit of Man making him more 'sweet' and more of a 'Christ-centred light bringer' – the 'genericness' of a compounds' etymology can now be seen in that the allocation here is close to, but not identical with, its representation in the Calvin cycle where it applied to working on Man's whole spirit as well. Chemical formulae from Genesis endnote[x].

***'1,3-Bisphosphoglyceric acid' has an etymology of 'phosphorous' ('the light bringer') and 'glycerine' ('sweetness').*** This compound is D-chiral; and thus, represents the working Holy Spirit, and there are two phosphate groups present. In the Calvin cycle it represented the Holy Spirit in His role working with the 'light produced' by two light bringers (Christ and the regenerated Man were the light bringers); but now 1,3-bisphosphoglyceric acid is the working Holy Spirit working on Man's original spirit (the '1') with the source of 'the three resources' (Christ's, the Holy Spirit and Man's regenerated spirit's image lives – the '3'): He now plants the Bible and the Holy Spirit (*'stars'* – the 'bisphospho') in Man – two more 'sweet' light-bringers (endnote[xi]).

***3-phosphoglycerate (D-chiral), 2-phosphoglycerate (D-chiral), phosphoenolpyruvate (non-chiral) and pyruvate (non-chiral).*** In consistency with the previous representations the etymology remains consistent but with the '3' representing working with Christ's, the Holy Spirit and regenerated Man's image lives; and the '2' Man's original spirit being made into a regenerated one. Please note the first two are, by D-chirality, what the working Holy Spirit works with and the etymology represents Man, whereas the second two are representing the Holy Spirit and Bible by etymology only – this is consistent with the concept of the latter two being independent lights ('permanent lodgers') with no *'kol'*, as opposed to Christ's gifts which are Man's, endnotes[vii, viii, ix].

***Genesis 1:14-19 (KJV): Before salvation: 14. And God said, Let there be lights in the firmament of the heaven to divide the day from the night; and let them be for <u>signs, and for seasons, and for days, and years:</u>***

***15. And let them be for lights in the firmament of the heaven to give light upon the earth: and it was so.***

**At salvation:** ***16. And God made two great lights; the greater light to rule the day, and the lesser light to rule the night: he made the stars also.***
***17. And God set them in the firmament of the heaven to give light upon the earth,***

**After salvation:** ***18. And to rule over the day and over the night, and to divide the light from the darkness: and God saw that it was good. 19. And the evening and the morning were the fourth day.***

**The Proof and the Pattern (**please see figure 11.1 for the allocation of the numbers used below)

(Please consider this order: signs ∞ seasons ∞ days ∞ years.)

[1] **Glucose:** ***God said let there be lights in the firmament of the heaven to divide the day from the night; and let them be for signs, and for seasons, and for days, and years*** – the working Holy Spirit will bring four lights and all the parallels represented by glucose to the whole world. The Holy Spirit – proclaiming Jesus as the light of the world – God the Father purposes this action.

[2] **ATP → ADP:** God's light to Man – *'ereb to boqer'* – resulting in:

[3] **Glucose 6-phopshate:** ***And let them be for lights in the firmament of the heaven to give light upon the earth: and it was so.*** The Holy Spirit presents specifically Christ as the light of life to individual Man – Jesus Christ is the light bringer.

[4] **Fructose 6-phosphate:** Man accepts salvation; the Holy Spirit has now been fruitful, represented by the name change.

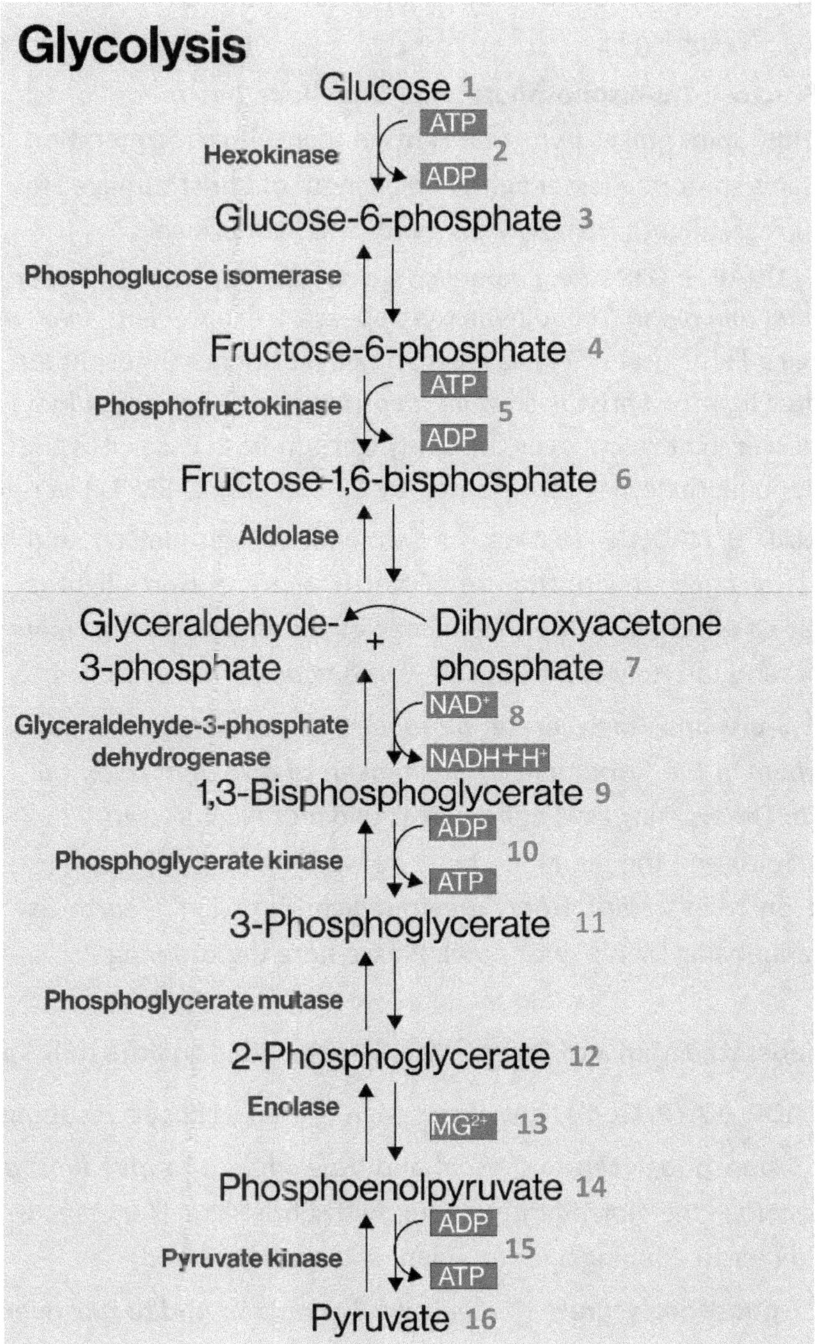


**Figure 11.1: Glycolysis; likkii/Shutterstock.com**

[5] **ATP → ADP:** More of God's light goes into Man – *'ereb to boqer'* – resulting in:

[6] **Fructose 1,6-bisphosphate: *And God made two great lights;*** the Holy Spirit makes two great lights in Man with a regeneration of Man's spirit (the lesser light) and an input of Christ's image life (the greater light); He now works with two light bringers.

[7] **DHAP → G3P: *the greater light to rule the day, and the lesser light to rule the night:*** The dihydroxy appears to represent 'two water bringing lights' (two 'O's) as a part of a light bringer (phosphate). This greater light of Christ is creating and ruling over goodness (day) and the lesser light ruling over the sin (night). In both cases it is the Holy Spirit doing the ruling represented by G3P. In effect, G3P is the ruling.

[8] **NAD → NADH + $H^+$:** Man is now a light bringer himself, and this is recognised with the use of NADH. More of God's light is about to enter Man (the indwellings of the Bible and the Holy Spirit) and the Holy Spirit Himself – *ereb to boqer* – resulting in:

[9] **1,3-bisphosphoglycerate: *he made the stars also. And God set them in the firmament of the heaven to give light upon the earth.*** The working Holy Spirit places two more lights (stars – the Holy Spirit and the Bible) in Man. The working Holy Spirit shines light on Man's spirit from 'the four lights in Man' (*'earth'* is metaphorising Man's spirit now). Please note the ordering:

Two great lights ∞ stars

**(Regenerated Man and Christ in Man) ∞ (the Bible and the Holy Spirit)**

[10] **ADP → ATP:** God's light enters Man – *'ereb to boqer'* resulting in:

[11] **3-phosphoglycerate ('*Pison*' and '*Havilah*'): *... to give light upon the earth*** – the Holy Spirit working with Christ-like likenesses to the world ('earth' flipping back to meaning the whole world).

[12] **2-phosphoglycerate ('*Gihon*' and '*Ethiopia*'):: *And to rule over the day*** – the Holy Spirit working with and presenting Christ's image and light internally to the spirit of Man and regenerating it.

[13] $Mg^{2+}$**:** The Holy Spirit is about to help draw more converts in like a magnet.

[14] **Phosphoenolpyruvate ('*Euphrates*'):** ***and over the night*** – the Holy Spirit will help Man to know good from evil and to help teach others about good and evil too in order to spiritually grow and evangelise, respectively (2 Timothy 3:16[58]). This will help Man to overcome darkness ('*night*') in himself and the world.

[15] **ADP → ATP:** God's light goes into Man – *'ereb to boqer'* – (whenever Man reads the Bible God's light goes into Man and the Holy Spirit is the 'light bringer') leading to:

[16] **Pyruvate ('*Hiddekel*'):** ***and to divide the light from the darkness.*** The Bible with the Holy Spirit helps Man differentiate right from wrong and help in the evangelisation of others.

Please note that there is a juxtaposition for the last two entities from the normal 'birthing sequence' order:

**Regenerated Man ∞ Christ in Man ∞ the Holy Spirit ∞ the Bible and the Holy Spirit**

**3-phosphoglycerate ∞ 2-phosphoglycerate ∞ phosphoenolpyruvate ∞ pyruvate**

The birthing sequence is an at-salvation description, but long-term after salvation and also before salvation, this order changes in that categories 3 and 4 juxtapose. This happens in Genesis 1:22.[59]

**fruitful ∞ multiply ∞ fill the waters ∞ fowl fill the earth**

(It also happens in Genesis 1:12).

---

[58] 2 Timothy 3:16 (KJV): ***All scripture is given by inspiration of God, and is profitable for doctrine, for reproof, for correction, for instruc tion in righteousness:***

[59] Genesis 1:22 (KJV): ***And God blessed them, saying, Be fruitful, and multiply, and fill the waters in the seas, and let fowl multiply in the earth.***

**Why the change in sequence?**

The original 'resurrection of Christ' can also be interpreted in the same two ways: Christ's instruction to *'receive the Holy Spirit'* may not be in the future but there and then (now moving it into 'third place' and not fourth) and the final Pentecost event is still paralleling the Holy Spirit and Bible; but that now moving into 'fourth place'. By this understanding, the second pattern is biblically confirmed. (It is a debated point in Christian theology about when the apostles actually received the Holy Spirit because of this supposed ambiguity – I think it is a deliberate dual-optioned design.)

This means, as well as the 'birthing sequence' or 'partum' sequence, there is also a, very similar 'antepartum', and an identical to that 'postpartum' sequence; and together all three these account for nearly all of the Genesis 1 and 2 text.

**Man ∞ Christ in Man ∞ the Holy Spirit ∞ the Holy Spirit+Bible**

I suggest the two sequences are to represent:

1. **Christ as a Man;** the-resurrection-of-Christ sequence; this is the 'mechanism' by which Man is saved (the blood sacrifice of a perfect life) and so the 'birthing sequence' follows this pattern in the KRM.
2. **Christ as God and a Man;** this is the modified sequence and represents the 'mechanism' by which Men initially come to salvation, how saved Man spiritually grow, and how they help in evangelising. Christ both as a man and Christ as God are needed for these actions.

A very clear Genesis-aligned pattern can be seen (almost word for word). So, just using word etymology, chemical formulae and the specific order of reactions in biochemistry, God can define salvation by Christ. Incredible!

# Chapter 12

# The Citric Acid Cycle

# (Krebs Cycle)

*Genesis-day-aligned biochemistry (V)*

The citric acid cycle is a series of enzyme-catalysed reactions that produce high-energy molecules in animals and plants to be used for ATP production and also form carbon dioxide.

## The Theory

The citric acid cycle parallels Day 5 in Genesis in remarkable ways with consistent metaphoric meaning with the previously studied processes.

The KRM book shows that whenever the word '*whole*' ('*kol*') is used it refers to an interaction – a two-way process, with both parties giving and receiving – but this makes for different 'counts' based on perspectives. The total number of likenesses described in Genesis 2:10-14 is ten. I have shown how, from the **working** Holy Spirit's perspective He makes all ten – making the interaction itself from God. From the working Holy Spirit's perspective He also uses ten in men when working on the salvation of other men; the two '*kol*' gifts are then Man's. I showed how this would make for 12 likenesses for the working Holy Spirit when considered as working before and after salvation (or with both Man and God); but as we have already seen, these water-based likenesses can be partitioned in several other ways too.

**Another bio-etymological role play**

Elements and biochemical compound names will, by their etymology, when viewed in a biblical context, reveal **the role** of an entity they represent within a certain aspect of the salvation process (often by extension of an established biblical metaphor). That entity will be one of, or a combination of, God the Father, the Lord Jesus Christ, the Holy Spirit, Man's original spirit, Man's regenerated spirit and the Bible: or any image life or likenesses given by God to Man – **what is the role? Who or what has that role? Where are we in the salvation timeline? Does chirality and the chemical formulae confirm it? (God, in effect, uses the same metaphors in the Bible and biochemical names.)**

**A detailed etymological cross-match**

I have shown that if I use very specific etymological meanings, then very specific perspectives of entities can be considered, and these can be cross-matched with the Genesis text through a chemical element to categories-within-a-biblical-metaphor way – by this method the Genesis text can contain a confirmation of very many chemical formulae. I stress that there is significant perspective in this process, I believe in time, and as the KRM is extended to cover more processes, then the 'rules' of this process will be tightened. Please also note that compounds (and sometimes chirality within them) seem to have their own time frame – applying to all of salvation or a specific period within it (these time periods are often associated with one particular river of Eden).

**The citric acid cycle – the actors**

***'Citrous' has etymology associated with a citron tree and possibly also a cedar tree.*** In the Bible the citron trees are 'goodly' trees (Leviticus

23:40[60]), and have an association with the Feast of Tabernacles which in turn, is associated with Christ giving 'living water'; and cedar is associated with splendour, durability and building temples (1 Kings 6:9[61]). Man's body is considered a temple for the indwelling God (1 Corinthians 6:19[62]). Christ, and the born-again Man are associated with '***good** gold*' in '*Pison*' (Genesis 2:11). This is consistent with 'citrate' or the 'goodly tree' representing Christ in Man but also the package that delivers this goodness – '*good gold*', '*bdellium*' and '*onyx stone*'.

***'Aconitate' has an etymology of 'poison from a plant' (aconitum plant) and a 'dart' or 'javelin'.*** In the Bible, sin in Man is associated with poison and also 'fiery darts' (Ephesians 6:16[63] and Deuteronomy 32:33[64]). This is consistent with 'aconitate' being a reference to Man's original spirit and the sin within it – sin to be removed as it is 'paid for'.

***Isocitrate: as with citrate above***, but please note 'iso' has a meaning of 'the same' and also note the use of the term 'the same' for the river '*Gihon*' in Genesis 2:13.

***'Alpha-ketoglutarate' (α-ketoglutarate) has the following etymology: 'glutamate' has a meaning of 'glue' or 'grain', 'ketone' derives from 'acetum' (vinegar) and 'alpha' means 'first'.*** This is consistent with alpha-ketoglutarate representing the sacrifice of Christ (vinegar) as the

---

[60] Leviticus 23:40 (KJV): ***And ye shall take you on the first day the boughs of goodly trees*** *(thought to be a citron tree)* ***branches of palm*** *(next page)* ***trees and the boughs of thick trees and willows of the brook; and ye shall rejoice before the LORD your God seven days.***

[61] 1 Kings 6:9 (KJV): ***So he built the house, and finished it; and covered the house with beams and boards of cedar.***

[62] 1 Corinthians 6:19 (KJV): ***What? know ye not that your body is the temple of the Holy Ghost which is in you, which ye have of God, and ye are not your own'***

[63] Ephesians 6:16 (KJV): ***Above all, taking the shield of faith, wherewith ye shall be able to quench all the fiery darts of the wicked.***

[64] Deuteronomy 32:33 (KJV): ***Their wine is the poison of dragons and the cruel venom of asps.***

glue that holds together; Christ is referred to as 'alpha'.[65] 'Grain' in the Bible is also a reference to a resurrection.[66] Thus, the use of 'α-ketoglutarate' is consistent with this being a reference to a resurrection of Man's spirit (a continued regeneration), and a carrying of a bond with Christ; in *'the body of Christ'*; or it may also be a reference to the 'cooperative' – being 'glued' together; and my position is it represents all three.

***'Oxalosuccinate' has an etymology of 'oxalic acid' (from 'oxalis', meaning 'acid from wood') and 'succinate', meaning the 'amber of a tree'.*** Wood is often a reference to Christ's crucifixion in the Bible. 'Amber of a tree' is almost a 'standard' Holy Spirit metaphor by now. Just as oxaloacetate was shown to represent Christ-like attributes or of Christ, oxalosuccinate represents the Holy Spirit carrying both his own and Christ-like attributes and some part of both of their image lives.

***'Succinic acid' has the same etymology as the 'succinate' part of 'oxalosuccinate'*** – representing anything 'from' the Holy Spirit.

***'Acetyl coenzyme A' and 'succinyl coenzyme A' have an etymology of 'enzyme' meaning 'leaven', 'acetyl' meaning 'vinegar', and 'succinate' meaning the 'amber of a tree'.*** The Bible's use of 'leaven' is associated with three 'measures' ('the three resources'); the structure of coenzyme A is of an adenine nucleotide with a third phosphate group attached. In line with other nucleotide use, it could be interpreted to be referring to the Holy Spirit working (ribose) with three 'light bringers' and specifically included is bringing Christ's work or gifts with it (adenine and acetyl), and the Holy Spirit's own work or gifts (succinyl).

---

[65] Revelation 22:13 (KJV): ***I am Alpha and Omega, the beginning and the end, the first and the last.***

[66] John 12:24 (KJV): ***Verily, verily, I say unto you, Except a corn of wheat fall into the ground and die, it abideth alone: but if it die, it bringeth forth much fruit.***

***'Fumarate' has an etymology associated with smoke from a plant***. Smoke is often likened to the Holy Spirit and His blessings, presence and power[67]. 'Fumarate' can thus, represent Man carrying these blessings of Holy Spirit and Christ-like spiritual growth in Himself.

**Malate, oxaloacetate** and **pyruvate** have already been covered.

**Carbon counts** (a recap of Chapter 2)

God, at salvation, gives Man five image lives and regenerates Man's spirit (Christ's image life, Christ and the Holy Spirit in Man's living waters, *'the body of Christ'* and *'the bride'* of Christ) – five in total ('God's five image lives given') and one continuing regeneration – six ('God's six image lives given').

Man's born-again spirit contains his original spirit, his regenerated spirit, his *'beast of the earth'*, and a presence in the *'the body of Christ'* and the *'bride'* of Christ – five in total ('Man's five image lives'); but Man gets more of Christ's image life – six ('Man's six image lives').

*'Pison'* is associated with three resources *'good gold'*, *'bdellium'* and *'onyx stone'* ('the three resources'), and three image lives are their sources ('the three image lives of *'Pison''*); it also acts with Man's spirit (*'whole land'*) to make four ('the four image lives of *'Pison''*).

**Oxygen counts**

God uses four lights on Man – Himself, Jesus Christ, the Holy Spirit and the Bible ('God's four lights given'), and Man has four lights – the lesser light, the greater light, the Bible and the Holy Spirit ('Man's four lights'). When both are considered together with the overlap removed, this adds up to six lights ('Man and God's six lights'). All of these may have the additional long-term addition of more of Christ's light: making 'God's five lights'; 'Man's five lights'; and 'Man and God's seven lights'.

---

[67] Revelation 15:8 (KJV): ***And the temple was filled with smoke from the glory of God, and from his power; and no man was able to enter into the temple, till the seven plagues of the seven angels were fulfilled.***

## The Proof and the Pattern

**Citric acid ($C_6H_80_7$) – citrate ($C_6H_50_7^{3-}$)** – a 'goodly tree'. **It is achiral** – Citric acid is thus Christ, the Holy Spirit and regenerated Man (**'*Pison*').**

**[1]** Citric acid gives **[2]** hydrogen ions at **[3]** physiological pH to become **[4]** citrate

**[1]** Saved Man as a 'goodly tree' gives **[2]** 'light' at **[3]** salvation **[4]** into the outside world to be 'pleasing to the sight of God'.

God uses 'God's six image lives given' in the conjugate acid (in making the 'goodly tree'); God makes or adds to 'Man's six image lives' in the conjugate base (the working 'goodly tree').

The conjugate acid represents what forms the 'goodly tree' and the goodly tree contains by etymology, 'the three resources' and is not just a 'one-likeness Pison' – but a 'three-likeness Pison' – the three likenesses are needed to 'live up to' its etymological name – $H_3$. As this includes 'Christ, the Holy Spirit and regenerated Man' it requires 'God's five waters to flow' – $H_8$ in total. The conjugate base containing $H_5$ reveals forward progresses and the external pointing likenesses that this river produces; thus the 'goodly tree' makes available for Man 'the five water likenesses of Pison' – $H_5$.

The electrons hover around the oxygen atoms in the conjugate base citrate in consistency with the Holy Spirit working with 'lights'. Chemical formulae from Genesis.[xii]

**Cis-aconitic acid ($C_6H_6O_6$) – cis-aconitate ($C_6H_3O_6^{3-}$) is achiral** – and represents Man in His sinful-nature cleaned by the 'indwelling package' and being cleaned by the process to become sin free ('***Havilah***').

**[1] Cis-aconitic acid** gives **[2]** hydrogen ions at **[3]** physiological pH to become **[4] cis-aconitate**.

**[1]** Man interacting with the Holy Spirit gets **[2]** 'lights' at **[3]** salvation **[4]** to purify Man's spirit and instantly remove sin.

God uses 'God's six image lives given' when purifying Man; God makes or adds to 'Man's six image lives' in the purification.

The 'purifying' Holy Spirit interacts with Man. Cis-aconitate is transient and disappears quickly as does Man's sin in the parallel. The chemical formula will reveal what Man has to do in order to 'live up to' his etymology and here it is 'to have sin removed'.

Man gives six aspects to making living waters flow $H_6$ (cis-aconitic acid has $H_6$). In order to be washed, aconitic acid (representing the interaction of the '*whole land*' *of Havilah*') must continually interact with the '*good gold*', '*bdellium*' and '*onyx stone*' – $H_3$. The conjugate acid has $H_6$ and the base $H_3$ in consistency with this.

The remaining negative charges are associated with the oxygen atoms, and this is consistent with the Holy Spirit working with 'lights' in Man. Chemical formulae from Genesis.[xiii]

(***'Gihon'*** **and** ***'Ethiopia'*****:** are represented by three compounds, each one representing a specific component of the events in '*Gihon*': isocitric acid; oxalosuccinic acid; and α-ketoglutaric acid – these represent – the 'goodly tree'; Christ and the Holy Spirit image life being given or used and then gluing this 'cooperative' together; or Man being 'glued to Christ' and in the '*body of Christ*'; all respectively.)

**Isocitric acid ($C_6H_8O_7$) – isocitrate ($C_6H_5O_7^{3-}$) is D-chiral (**unlike citric acid and citrate**)** – and so, is the working Holy Spirit by D-chirality working at adding Christ's image life to Man. By the use of '*iso*' this is still the same 'goodly tree' as found with 'citric acid' – containing the image lives of 'the three resources'; but now Man's whole original spirit is being changed and regenerated and not just purified. The hydrogen count in the acid-base split is the same as with citric acid and citrate but with '*Gihon*' now containing 'the three resources' ($H_3$) and not '*Pison*' and 'God's five water likenesses' ($H_5$) still applying. By D-chirality isocitric acid represents the **working** Holy Spirit giving to Man and not just a goodly tree 'shining out of' Man as found with citric acid and

citrate. For the base citrate, the Holy Spirit uses the two likenesses of Gihon (*'Gihon'* and *'Ethiopia'*) and with 'the three resources' added (Gihon now also has these by the etymology of isocitrate) – isocitrate has $H_5$ in consistency. The chirality adds a second layer in that both the 'goodly tree' and the working Holy Spirit working with the 'goodly tree' have identical individually-confirmatory chemical element counts.
**[1]** Isocitric acid gives **[2]** hydrogen ions at **[3]** physiological pH to become **[4]** isocitrate.

**[1]** The Holy Spirit gives **[2]** image life and 'light' to Man to form the 'goodly tree' package **[3]** at salvation [**4]** and form more 'goodly tree'.

The remaining negative charges are associated with the oxygen atoms, and this is consistent with the Holy Spirit working with 'lights' in Man. Chemical formulae from Genesis[xiv].

**Oxalosuccinic acid ($C_6H_6O_7$) – oxalosuccinate ($C_6H_3O_7^{3-}$) is L-chiral** – the image life (L-chirality) and light of Christ being carried by the Holy Spirit (succinate). Please note that this compound quickly becomes α-ketoglutarate – and Christ's image life and light is quickly given to Man in the parallel and, in a way, this is all a part of the same reaction that forms α-ketoglutarate. As it is **L-chiral**, it represents specifically, the Lord Jesus Christ – this is the Holy Spirit carrying Christ's image life and light as well as His own. Compare this to **achiral** oxaloacetate – Man's own self, carrying the same, Christ-like light (but Man now), and **L-chiral** malate – man in the *'body of Christ'* – a consistent interpretation of L-chirality. The remaining negative charge is associated with the oxygen atoms – the Holy Spirit working with 'lights' in Man ('Man and God's seven lights'). Formulae from Genesis[xv].

[1] Oxalosuccinic acid gives [2] hydrogen ions at [3] physiological pH to become [4] oxalosuccinate.

[1] The Holy Spirit carrying Christ's light gives [2] 'Christ's image life and light' at [3] salvation [4] to make Man contain Christ's image life and live in the *'body of Christ'*.

**A-ketoglutaric acid ($C_5H_6O_5$) – α-ketoglutarate ($C_5H_4O_5^{-2}$) is achiral (*'Ethiopia'*)** – this is the original spirit of Man 'interacting' with the 'cooperative' (this interaction is a representation of *'the whole ('kol') land of Ethiopia'*). Man contributes six aspects to make living waters flow (α-ketoglutaric acid has $H_6$). Four likenesses result $H_4$ (*'Pison'*, *'Havilah'*, *'Gihon'* and *'Ethiopia'*) – these are all the 'active' likenesses associated with *'land'* (*'eres'*). The oxygen charges parallel the Holy Spirit working with 'lights' in Man.

[1] A-ketoglutaric acid gives [2] hydrogen ions at [3] physiological pH to become [4] α-ketoglutarate.

[1] Man's gives and receives [2] 'light' in an interaction with the Holy Spirit at [3] salvation [4] to become regenerated and to form a bond with Christ – the two together. Chemical formulae from Genesis.[xvi]

**Succinyl CoA** – is the 'nucleotidal way' of describing what the Holy Spirit carries (Himself and the Bible) and presents them to Man – the working Holy Spirit presents what the Holy Spirit carries to Man.

**Succinic acid ($C_4H_6O_4$) – succinate ($C_4H_4O_4^{-2}$) is achiral (*'Hiddekel'*) –** Man carrying the Holy Spirit (succinic). Man contributes six aspects to make living waters flow (succinic acid has $H_6$). Up to this stage in the 'rivers' timeline, Man receives from the Holy Spirit *'Bdellium'*, the Holy Spirit's part as 'bdellium' in the 'cooperative', *'Hiddekel'* and *'Assyria'* – $H_4$; succinate has $H_4$ in consistency. The negative charges are associated with oxygen atoms, and this is consistent with the Holy Spirit working with 'lights' in Man. Chemical formulae from Genesis.[xvii]

It should be noted that both *'Hiddekel'* and *'Euphrates'* are not covered by *'kol'* – and so are not a 'two way interaction', but the analogous *'winged fowl'* in Genesis 1:22 *'kol'* is used (translated as 'every'). This is because the Genesis 1:22 is at salvation (the moment they are given) and aligned with the word 'created' (this is the day that they are created from the Holy Spirit) whereas the rivers of Eden are long term, hence, *'Hiddekel'* and *'Euphrates'* are a one-way process.

**Fumaric acid ($C_4H_4O_4$) – fumarate ($C_4H_2O_4^{-2}$) is achiral (‘*Euphrates*’):** The regenerated Man carrying a blessing from God to be fruitful and multiply (the whole spirit also has this fruitful blessing). Regenerated Man contributes four aspects to making living waters flow (fumaric acid has $H_4$). Regenerated Man has two likenesses (‘*onyx stone*’ and ‘*onyx stone*’ in the cooperative) – fumerate has $H_2$.

Euphrates (meaning ‘fruitful’) is not like the other rivers, in that no interaction or participants are described. This is consistent with it representing a specific blessing enclosed within the others, but something that comes together to be fruitful (spiritually add to the ‘*body of Christ*’). This fruitfulness is also found in ‘malate’ and ‘oxaloacetate’ representing a part in the ‘*body of Christ*’ and Christ-like attributes. **All fruitfulness comes by the working Holy Spirit, both internally and externally and by this understanding of ‘fruitful’ a cyclical nature with ‘fruitful’ being either end of the ‘one true pattern’ becomes apparent – one Christ centred and one Holy Spirit centred**.

The blessing to be fruitful is both for Man in himself and also for him to work with other men – the blessing is needed for Man to become a ‘malate’ and in effect to give off ‘oxaloacetate’. The remaining negative charge is associated with the oxygen atoms, and this is consistent with the Holy Spirit working with ‘lights’ in Man.

[1] Fumaric acid gives [2] hydrogen ions at [3] physiological pH to become [4] fumarate.

[1] Man gives out [2] ‘light’ via the Holy Spirit at and [3] after salvation [4] to use his blessed ability to aid the Holy Spirit in being ‘*fruitful*’.

Chemical formulae from Genesis.[xviii]

**Malate, oxaloacetate** and **pyruvate**: have been previously covered and their representation remains consistent. Please remember that these representations are for roles only (the latter two with no chirality), and pyruvate and phosphoenolpyruvate in particular, could, in different circumstances, both be representative of the Holy Spirit.

## **Krebs Cycle** (and the rivers of Eden)

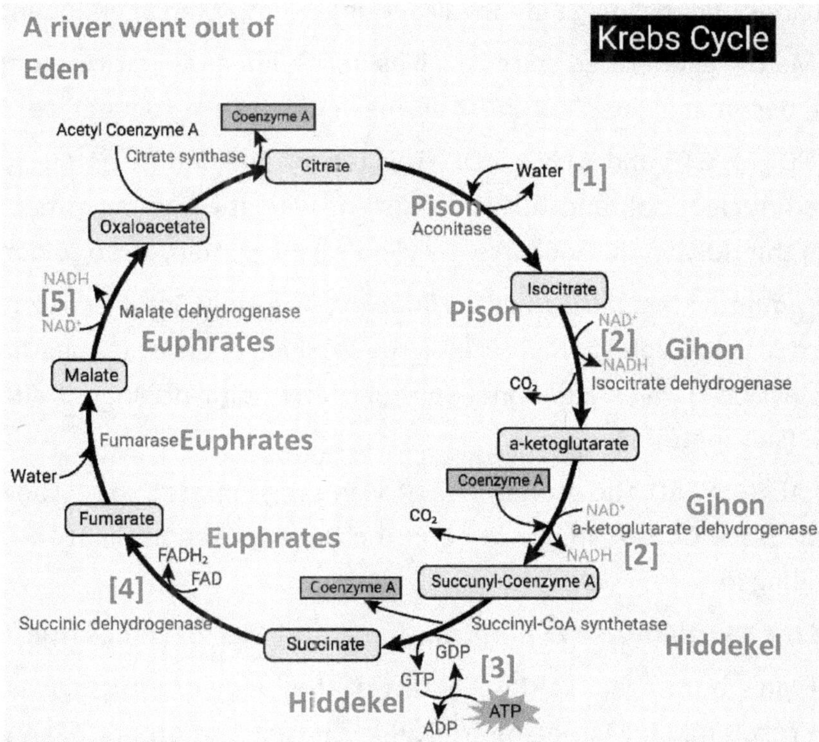


**Figure 12.1: Krebs cycle; Jaya Bharathi/Shutterstock.com**

***A river flowed out of Eden and watered the garden***. Man's living waters (malate and oxaloacetate) flow out of Man and interact ('whole land') with the indwelling 'package' (Jesus Christ and the Holy Spirit, bringing Christ's sacrifice – acetyl CoA) and the Bible (pyruvate).These participants will also feature ate the end of the cycle.

***And became into four heads***.

***The name of the first is Pison:*** Jesus Christ ('*good gold*'), the Holy Spirit ('*bdellium*') and the regenerated aspect of Man's spirit ('*onyx stone*'); together form the goodly tree (citric acid); Man's spirit is washed clean of sin (aconitic acid). Man gives off Christ-like attributes 'pleasant to the sight' of God.

[1] Living waters are used and produced:

***The name of the second is Gihon***: the working Holy Spirit (isocitrate) works with Christ's image life and pours the light of Christ onto Man and adds Christ's image life (oxalosuccinate) into Man in abundance:

[2] Man's regenerated spirit receives image life and light resulting in: Man becoming 'glued' to Christ or one in Christ (α-ketoglutarate).

[3] GDP → GTP and ATP → ADP. This is now guanine (the Holy Spirit-based nucleotide), and this is consistent with the coming gifts being from the Holy Spirit. Concurrently, God's light → Man, *'ereb to boqer'*.

***The name of the third river is Hiddekel***: The Holy Spirit carrying the Word of God (Bible) and preaching the sacrifice of Christ (altogether as succinyl CoA) gives additional spiritual Holy Spirit-originated gifts to Man (succinate).

[4] $FADH_2$ → FAD, the cooperative of Man's regenerated spirit, the Holy Spirit, Jesus Christ's image life and the Bible give more light to Man, resulting in:

***And the fourth river is Euphrates:*** (fumarate) Man becomes fruitful.[68]

[5] Man's born-again spirit as a part of the *'body of Christ'* gives out light resulting in: Man being fruitful in himself and others. Man goes through the cycle, doing the same for his own spirit and others. (The Bible also often follows a 'one true pattern' with further instructions to be *'fruitful and multiply'*.[67] Man's and other Men's own 'malate[s]' and their 'oxaloacetate[s]' (part of the *'body of Christ'*) working with pyruvate (the Holy Spirit and the Bible) help them to spiritually grow and to bring others to Christ (*"fruitful and multiply"*)).

With careful etymological considerations, viewed in a biblical context, the citric acid cycle adheres to the Genesis text; and, I suggest, this shared design must mean a shared designer.

---

[68] Genesis 1:22 (KJV): ***And God blessed them, saying, Be fruitful, and multiply, and fill the waters in the seas, and let fowl multiply in the earth.***

# Chapter 13

# Codons for Christ

*Digital and analogue*

## The Theory: Codons and Protein Synthesis

DNA and RNA behave a bit like five-letter digital code: the A, G, C, T and U act like the 0s and 1s – either on or off with nothing in between. But once we know these letters also have metaphor-holding names this adds a whole new layer of variable meaning – analogue meaning. A (the Lord Jesus Christ), G (the Holy Spirit), C (the Word of God or Bible and the Holy Spirit together), T (the spirit of a Man or Christ as a man) and U (the regenerated spirit of Man).

RNA is all about making proteins; they are used in just about everything (growth, repair, structure, hormones and enzymes). DNA holds the code for making proteins, and during this process, just like in DNA replication, it splits into its two strands. Messenger RNA then copies one strand but copies it as a complementary pair (called 'transcription'). The DNA strand that serves as the 'pattern' or 'template' for this copy is called the 'template strand'. The RNA strand is thus, an exact copy of the other strand of DNA, as they are both complementary strands to the template strand, with one exception in that thymine is replaced with uracil in RNA. The non-template strand of the DNA is called the 'coding strand', and both this and the similar, except for thymine versus uracil, RNA strand have the 'actual code' used in protein synthesis (called 'translocation'). Much emphasis is placed on these two coding strands in the study of DNA, this study will focus on mRNA codons but one can be deduced from the other.

The unit of currency in these coding processes is a group of three nucleotides, or 'three letters'. These are collectively called a 'codon' – each codon of three nucleotides codes for a specific amino acid, and multiple codons will sequentially code for peptides and proteins.

**What represents what?**

***'Ribosome' has an etymology of 'ribose' (the Holy Spirit) and 'some' ('body')***; hence, 'ribosomes' are organelles that make peptides and proteins, and by etymology, could be considered to represent the Holy Spirit working in the *'body of Christ'* (all Christians).

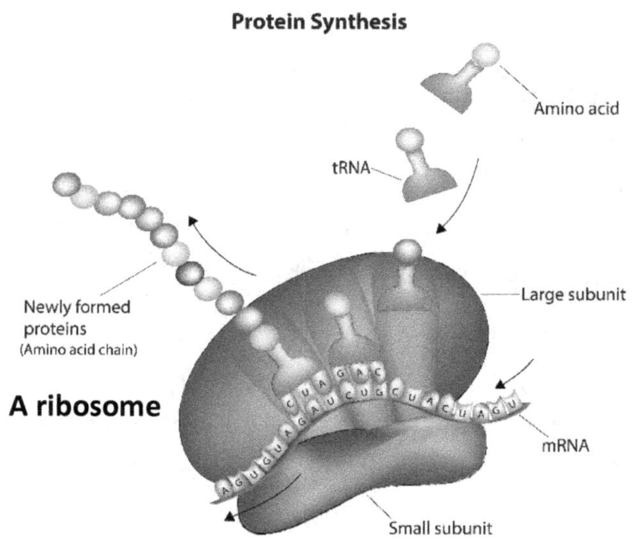


**Figure 13.1: Protein synthesis; contributor: KKT Madhusnaka/Shutterstock.com**

The messenger or mRNA and transfer or tRNA, together with ribosomal or rRNA, represent working in saved Man – and the amino acids represent changes they produce in Man. More specifically, **mRNA is what the Holy Spirit brings from/for God (from the DNA template strand), mRNA, tRNA and rRNA represent what is in Man, and I will**

**show that the etymology of the amino acid they yield represents some very specific gifts given to Man in this post-salvation process.**

Each amino acid is carried by the tRNA to the mRNA – the tRNA carries these amino acids on its 'back'; and when it complementarily pairs with mRNA, as shown in figure 13.1, it brings the amino acid with it. Successive tRNA strands then travel to the mRNA in the ribosome and sequentially bring more amino acids that join together to form a protein, and in each case the codon defines which amino acid is used.

There are 64 possible options when you have five characters (A, G, C, T and U) in sets of three (there can be no blanks). There are only 20 common amino acids; so, as there are 64 options for codon sequences, some amino acids are coded by more than one codon. Three codons are called 'stop codons', as they end the protein-making process, and there is only one codon that is a 'start' codon that begins it (methionine). **If you complementarily pair to a codon, you get an anticodon** – thus, tRNA has the anticodon to the codon in mRNA in order for it to 'fit into' the mRNA – codons can represent what God sends out to Man – anticodons will represent what God works with in Man, but all codons and anticodons are working together in Man, a transaction. Thus, A in mRNA pairs with U in tRNA, and C in mRNA pairs with G in tRNA, or the same in reverse – U in mRNA pairs with A in tRNA, and G in mRNA pairs with C in tRNA. It is really the interaction between the two that counts – the Holy Spirit 'working' in Man.

The amino acid methionine is coded by codon AUG in mRNA: AUG in mRNA pairs with UAC in tRNA (as this process is with complementary pairing) – you will note that 'UAC' is the same order as the first three categories of the 'birthing sequence', and if you factor in the 'Holy Spirit alone' (the fourth category) as represented by the next nitrogenous base (this is called position 37 which we will come to):

Regenerated Man ∞ Christ in Man ∞ the Bible ∞ the Holy Spirit

U ∞ A ∞ C ∞ a variable base at position 39

So, when proteins are made, it is almost always initiated by methionine, as it is the only start codon. This could be considered as Jesus Christ, Man's regenerated spirit and the Holy Spirit (AUG) working in Man in the 'birthing sequence' order (UAC), which is the order of salvation in Man as found throughout the KRM in Genesis and in biochemistry – the proteins formed representing Christ by chirality.

***'Methionine' has an etymology from 'methyl' and 'thio' ('sulphur')***; hence this could, as previously discussed, be interpreted to mean Man washed clean by the cross. This etymology could represent methionine (or the birthing sequence) as the start of all spiritual growth in Christ.

The etymology of some other amino acids is also interesting and salvation-gifts mimicking (please note that this is all after salvation):

**Alanine ($C_3H_7NO_2$):** this may be a derivative of 'aldehyde', but also '*alan*' has a Celtic history of a meaning of a '*little rock*'. Jesus is referred to as a 'rock', and by extension, the '*body of Christ*' can be too. (Maybe in Matthew 16:18,[69] the '*rock*' means Peter and Christ.)

**Arginine ($C_6H_{14}N_4O_2$):** '*bright shining white*' or '*silver*'.

**Asparagine ($C_4H_8N_2O_3$):** '*sprout*' or '*shoot*' (Christ is referred to as a 'shoot', 'branch', 'vine' or 'sprout'.)

**Aspartic acid ($C_4H_7NO_4$):** as with 'asparagine', but now an emphasis on '*acid*' – a 'dealer in light' in the KRM.

**Cysteine ($C_3H_7NO_2S$):** a '*bladder*' or '*sac*' – or '*containing water*'.

**Glutamic acid ($C_5H_9NO_4$):** from '*gluten*' or '*grain*' and also '*glue*', an 'acid' – a dealer in light (acid), a resurrection (grain – 'the falling and dying of wheat' of John 12:24) and a binder together (glue) in the KRM.

**Glutamine ($C_5H_{10}N_2O_3$):** as above: '*gluten*' or '*grain*' and also '*glue*'.

---

69 Matthew 16:18 (KJV): ***And I say also unto thee, That thou art Peter, and upon this rock I will build my church; and the gates of hell shall not prevail against it.***

**Glycine ($C_2H_5NO_2$):** *'sweet'*, **but please note glycine is the only amino acid here that is not L-chiral,** thus glycine could be interpreted to be a *'sweet tasting'* element to Man's original spirit or the Holy Spirit that is not part of the *'body of Christ'*– a very specific reason for its achirality?

**Histidine ($C_6H_9N_3O_2$):** *'tissue'* or *'weave together'*; this could be interpreted to be spiritual entities that are being weaved together.

**Isoleucine ($C_6H_{13}NO_2$):** *'iso'* – the same; *'leucine'* – *'white'* or *'bright'*.

**Leucine ($C_6H_{13}NO_2$):** *'white'* or *'bright'*.

**Lysine ($C_6H_{14}N_2O_2$):** *'to break down'* or *'loosening'*.

**Methionine ($C_5H_{11}NO_2S$)**: as above.

**Phenylalanine ($C_9H_{11}NO_2$)**: *'alanine'* with *'phenyl'; 'phenyl'* has an etymology of *'to appear'* or *'to shine'*. Together with 'alanine', this could be interpreted as an appearing *'shining little rock'*.

**Proline ($C_9H_{13}NO_2$):** this gets its name from 'pyrrole', which has an etymology of *'fire'* and *'oil'* – both are established biblical metaphors for the Holy Spirit and the Bible.

**Serine ($C_3H_7NO_3$):** *'silk'* – The Bible describes a link between 'silk' and a 'bride', silk adorning a bride.

**Threonine ($C_4H_9NO_3$):** has etymology associated with the sugar threose, which is named from a 'rearrangement' of the word 'erythrose'; it was used to differentiate between the two as threose is the diastereomer (has the same molecular formula) of erythrose. 'Erythrose' means *'red'* or *'blood'* or perhaps the *'red sugar'*.

**Tryptophan ($C_{11}H_{12}N_2O_2$):** *'to digest'* or to *'rub against'* (from *'tryptin'*) and to *'appear'* (from *'phan'*).

**Tyrosine ($C_9H_{11}NO_3$):** *'cheese'*. I take the position that there is a link with cheese as fermented milk – a term associated with nurturing and food moving from liquid to solid. 'Cheese' has an etymology associated with 'fermented thing' or 'to make sour' (make acidic).

**Valine ($C_6H_{14}NO_2$):** *'to be healthy'* and *'strong'*.

**The Proof: General Coding Stories** (the KRM gives the representation but we supply the rest of the story)

If we take a general look at some of the codons and the specific amino acids they code for (I am again simplifying some of this) then:

The CCC codon in mRNA codes for the amino acid proline ('fire' and 'oil' – both are established biblical metaphors for the Holy Spirit). This pairs with the GGG anticodon in tRNA. In the KRM, this can be interpreted to mean that the working Holy Spirit (the ribose in mRNA and tRNA) works with the Word of God (CCC in mRNA) and He specifically gives Holy Spirit image-and-likeness gifts (GGG in tRNA) to Man, which are metaphorised as changing Man (by the etymology of proline), with Man specifically now gaining more 'fire and oil' (proline) – this matches with the parable of the ten virgins (Matthew 25:1-13).

Please remember the structure of RNA is just like that of DNA except that there are five and not four oxygen atoms – $O_5$. Joined nucleotides always tell the story from a beginning to an endpoint, but with RNA there is an emphasis on more of Christ's light added to Man.

The tRNA code is like the DNA template code (except thymine vs uracil), and if you consider the whole process, this in a way, shows how all gifts originate from God (the template code), and specifically how the born-again Man (uracil) is from Jesus as a man ('T' in the God strand). This is because 'T' in DNA transcribes to 'A' in mRNA (Jesus as God in Man) and then pairs to 'U' in tRNA. The whole story when combined with etymology makes for a neat New Testament summary, and there are many other obvious parallels too.

The codon AAA (Jesus Christ as God) in mRNA has an anticodon of UUU (the regenerated Man) in tRNA. This AAA codon codes for lysine (*to break down*) – this is consistent with Jesus Christ (AAA in mRNA) breaking down Man's old sinful spirit and regenerating it in the process (UUU in tRNA). This can also be interpreted to mean more and more of

Christ's light goes into the regenerated Man – Man becomes more Christ-like – **the etymology provides the representation, but we fill in the blanks (often verbs)**.

Likewise in reverse: the codon UUU in mRNA couples with AAA in tRNA; and this can be interpreted to mean the regenerated Man is placed in Christ – in the *'body of Christ'*. UUU codes for the amino acid phenylalanine – the *'shining little rock'* – and this can also be looked upon as a term for thriving Christians 'shining' in the *'body of Christ'*.

Likewise, when the codon GGG in mRNA couples with CCC in tRNA (the Holy Spirit makes in Man); and this can be interpreted to mean the Holy Spirit *'made to grow'* the indwelling Bible and Himself together in Man (this is called the *'tree of life'* in Genesis 2:9). GGG codes for glycine: the 'sweet smelling' to God; so, this is not 'sweet smelling' by Christ's image life or likenesses; and this is consistent with this referring to Man's original spirit changed by the Bible and the Holy Spirit (the extra 'O' in ribose can then represent the Holy Spirit).

Please also note that tRNA is a complicated compound and contains variable amounts and types of nucleotides (on average about 70 to 90 and in a sort of cross formation); they also may change during the protein synthesis process. I would class mRNA and tRNA as representing the complicated interaction between Man and God after salvation. For instance, position 37 for methionine (U, A and C are position 34, 35 and 36) is often modified adenine or guanine and could represent Man as a modified representation of Christ or the Holy Spirit. (Please note the way the second digits roughly match with Genesis Days 4 to 7 when 'one true pattern' events are considered.)

**Essentially Christ**

Amino acids are divided into essential, conditionally essential and non-essential categories. Essential ones cannot be made by the body, so it is 'essential' for them to be in the diet. A review of the amino acids above with regard to their 'essentiality' and etymological meaning shows consistent KRM to real-physical-life correlations in their 'essentiality'. Just as the physical body of Man can produce non-essential amino acids, being in the *'body of Christ'* (saved) gives you certain gifts. Just as certain other amino acids need to be ingested, so, in parallel, more 'light' has to be given to saved Man if he is to get other spiritual gifts and to spiritually grow. There is much more on this to come.

**Wobbling *'light'* – image then likeness**

If we look at the codons that code for glycine, as we have already seen above, it is GGG; but so do GGC, GGU and GGA. The third base seems irrelevant, and any of the nitrogenous bases will suffice. This is because, usually, only the first two bases are essential; the third base can vary and may even also have non-standard pairing (sometimes even with different bases completely, like inosine[70]). This process is called 'wobbling' and is thought to give some needed 'slack' to the system, but I will show that all of it is very precise representation. And please note that the non-standard pairing (G with U), or pairing with a modified adenosine (inosine) and a modified guanine (queuosine), can still be representing salvation dependent on their own etymologies (*inosine* has etymology associated with 'muscles' and 'sinews', and these are symbolic of strength and restoration in the Bible). I will keep this study simple and not consider non-standard pairing.

The whole of the KRM is in an image-to-likeness format; we discussed that even chemical formulae appear to adhere to this (adenine $C_5H_5O_5$ – carbon and hydrogen images, and oxygen a likeness – with hydrogen sometimes a likeness). I believe this is what is happening with the third base variation. I suggest that the third base is

[70] Inosine – a naturally occurring purine nucleotide used in the various metabolic processes and used as a precursor for adenine and guanine. (If you remember, both adenine and guanine are purines – hence purines appear to be metaphorising with God and not Man.)

specifically, **a reference to category three of 'one true pattern' (a Holy Spirit-working likeness).** The first two bases are images **again as found in 'one true pattern', the first going out of Man and the second going into the now-indwelt Man**. Thus, GG (the Holy Spirit coming out of and going into Man) working with four Holy Spirit-worked-on likenesses from (U, A, C and G) will make Man 'sweet smelling' in the eyes of God (glycine). Thus, AA (Jesus Christ in and out of Man) and the Holy Spirit working with G and A will break down Man's sinful nature (lysine).

This becomes a way of numbering the participants and a direction of flow within a process, the 'wobbling' is needed as that is the number of 'likeness interactions' involved in the complex 'spiritual transaction' it represents. It will also help to explain why sometimes multiple codons code for the same amino acid, this has a misnomer of 'degeneracy', as I will show the number of codons shows pattern and is also predictable from the rivers of Eden passage. This comes from the previously mentioned separation of spiritual transactions and participants ('deliverers' and 'influencers'). This makes for a subdivision within the ten living water likenesses: living water interactions (seven) and deliverers delivering deliveries from custom sources (three). All will be represented by 'H' and I will list them later.

It may also be that the sequential etymological meaning of amino acids in peptides and proteins tell a 'story' that links with the protein's name or shows pattern in some other way. I believe that each amino acid may well need several Genesis-aligned books of their own!

I will also show that for all amino acids the Genesis text **can** even be interpreted to be aligned with their chemical structure; I will review a very specific 'journey' for each of the 20 common amino acids, and these will all link together and give a biblical reason for the presence of essential, non-essential and 'in-between' amino acids.

As always, the etymological meaning will be very significant: Does the etymology represent a Jesus Christ-like trait and his gifts to Man, or does it represent a role of the Holy Spirit and his gifts? Does the etymology apply to both? Does the etymology represent a specific event, characteristic, status, a specific Eden river or all Eden rivers? And most importantly of all does the etymology **select** different categories of 'one true pattern'.

Please note that rRNA within the KRM is at or long term after salvation; and the L-chirality of amino acids can represent *'the body of Christ'*, a *'bride'* for Christ or any Christ-like attribute.

**Putting things into perspective** (*'fruitful'*, *'Pison'* and 'alanine')

I will begin with methionine and alanine, and for alanine, the etymology represents Man's role as being a building block (a 'little rock') in the church. The church is the collective of all Christians often viewed as a community or in an evangelistic sense. The codons in mRNA (from God) for alanine are GC (U, A, C and G); the GC pairs with CG in tRNA (in Man). This can be recognising the saved Man using the Holy Spirit (G) and the Bible (C) as the entities needed in the 'transactions' to evangelise in order to be a part of the church. We know that category one in 'one true pattern' is an 'output' – an 'increase' from saved Man' – alanine's etymology fits that description.

For the same 'category-one' representation, oxaloacetate in malate synthesis took the perspective of traits from Christ; citrate in the Krebs cycle represented this role as three entities of a 'goodly tree'; and ribulose 1,5-bisphosphate in the Calvin cycle took the position of representing the working Holy Spirit. But all of them represent the first part of the Christ-centred category one of 'one true pattern'.

Once it is known that alanine is similar to *'Pison'* of Genesis 2:11 and *'fruitful'* of Genesis 1:28, then we can consider if the chemical formula of alanine is predictable as found with DNA. I am going to show how the etymological meaning of all the amino acids will have meanings that: **apply** to just Christ; **apply** to just the Holy Spirit; **can apply** to both; or **must apply** to both. But first something unifying and huge:

**DNA and RNA manifest 'one true pattern'**

I now want to point out a monumental and unifying aspect of DNA and RNA. The first two Christ-centred likenesses of Man are: *'fruitful'* and *'multiply'* and these can both be internal or external: becoming more Christ like yourself or making others more Christ like.

> ***2 Corinthians 4:6** (KJV)**: For God, who commanded the light to shine out of darkness (1), hath shined in our hearts (2), to <u>give</u> the light of the knowledge of the glory of God (3) in the face of Jesus Christ (4).***

If we go back to 2 Corinthians 4:6 we see that: saved Man can then be included as one who shines *'light'* on the whole world for God by this external meaning of his *'fruitful and multiply'* (1); an indwelt Christ shines light within the saved (2); the Holy Spirit and the Bible give knowledge (3); the use of *'face'* (*'prosopon'*) can mean a 'presence' and hence satisfy a *'fruitful'* Man being Christ-like (4). A 'one true pattern'.

Please also note, from a *'light'* perspective, the *'shine out of the darkness'* can be 'something for the process (all men) and also the individual saved Man', as the *'light'* in category 1 is the same entity that gives the *'knowledge'* of category 2.

In the rivers of Eden passage the *'gold'* (Christ), *'bdellium'* (the Holy Spirit) and *'onyx'* (saved Man) were found to be the 'deliverers or influencers' carrying gifts for their respective entities in living water production as opposed to the seven interactions themselves, this is a different (and very significant) ordering to 'one true pattern'. If we look at a base pair of DNA (Figure 2.2), then from left to right: Christ is a light bringer (phosphate); ribose is the Holy Spirit; and the two joined nitrogenous bases can be considered as the spiritual interaction in Man (the first base as a source and the second a result, in the general DNA 'story'). But when viewed from the 'other direction', the source can also be a likeness going out of Man to the whole world, born-again Man playing his role for God (just like category 1 of 'one true pattern') and the 'result-base' can still be spiritual changes going into Man or already in the indwelt Man but they are now a source. The paired bases then become category 1 and 2 of 'one true pattern'.

**Phosphate** (a); **ribose** (b); **base** – the sources coming out of Man (c) and (1): **base** – the results in Man (2); **ribose** (3); **phosphate** – Man as a light bringer (4).

A, b and c are consistent in order and representation with Jesus Christ ('*gold*'), the Holy Spirit ('*bdellium*') and Man ('*onyx*') as found in Genesis 2:11-12 and Genesis 2:19; followed by, and overlapping with, a 'one true pattern'. These two sequences together are exactly consistent with the ordering in Genesis 2:19-20 (remember that '*beast of the earth*' are in the '*body of Christ*' and '*fowl of the air*' represents the Holy Spirit in that passage). A, b and c can represent just a presence in Man or actual work on Man (just like '*gold*', '*bdellium*' and '*onyx*'), and simultaneously, be something for the whole process with Man contributing. Thus:

**'The overall process'** (a summary term that will be used several times)

Thus, mRNA to tRNA coupling carries 'one true pattern' preceded by a biblically defined Christ, the Holy Spirit and born-again-Man order. This is going to be the core of understanding a major biblical alignment.

For mRNA: God (the DNA template) sends forth Christ, the Holy Spirit and saved Man to the whole world but he also uses the same three for spiritual growth in the individual saved Man (something for the process and something for the individual; something external and something internal; evangelising and spiritual growth; and the *kosmos* and the *ruah*).

Thus, concurrently, the phosphate group, ribose and nitrogenous base (representing 'good gold', '*bdellium*' and '*onyx*') can be 'read' as a presence in category 1 and be 'forming' category 2, as found in the rivers of Eden passage (all just by nucleotide structure).

Please note that the working Holy Spirit selects from any component of what: phosphate ($PO_4$); ribose ($C_5H_{10}O_5$); adenine ($C_5H_5N_5$); guanine ($C_5H_5N_5O$); thymine ($C_5H_6N_2O_2$); uracil ($C_4H_4N_2O_2$); cytosine ($C_4H_5N_3O$) represent in order to carry **out His work and the chemical formulae of the amino acids will reveal this selection** (the chemical formula has consistent representation throughout).

Please note again how the terms messenger, transfer and ribosomal are prefectly suited and chronological in defining the Holy Spirit's particular roles. A 'messenger' from God to 'transfer' spiritual gifts to Man who is living in the '*body of Christ*' (please always appreciate that likenesses become images when in the '*body*', in category one).

**More nuance to the rivers of Eden and likenesses**

Some 'one true patterns' have extra information about Christ in the first two categories – a Christ-expanded 'one true pattern' and this leads to a 'pattern of six'; for example, in Genesis 1:28:

> **Be fruitful (1a),** and (sub category: multiply (1b)), and **replenish the earth (2a)**, and (sub category: subdue it (2b)): and have dominion over the **fish of the sea (3)**, and **over the fowl of the air (4),** and over every living thing that moveth (the long-term command for growth).

{Standalone numbers will refer to these categories from here on.} Because broad terms are used, some of these can apply to both the Holy Spirit and Jesus Christ; and similarities exist in that '1a' and '1b' are very similar in nature to '2a' and '2b' (one going out of Man and one going into Man); '2a; and '2b' are similar to '3' and '4' (both are additions and purifications); and '2b' can be a summary of '2a' and '2b'; and '1b' can be a summary of '2a' and '2b'. Then, by this understanding each set of six can be split into two further 'nested' sets of four with categories '2a' and '2b' overlapping. **This overlapping also emphasises the shared works of Jesus Christ and the Holy Spirit** (in effect you can read each pattern simultaneously and then they are working together).

**Thus, the category 2a and 2b, although Christ centred, are representing the Holy Spirit's work too.**

Fruitful ∞ multiply ∞ replenish ∞ subdue

Replenish ∞ subdue ∞ fish of the sea ∞ fowl of the air

In the KRM, mRNA codons pairing with tRNA anticodons represent a spiritual transaction and the participants are the nitrogenous bases.

> **Genesis 1:28 (KJV):** ***And God blessed them, and God said unto them, Be fruitful*** *[N1]*, ***and multiply*** *[N2]*, ***and replenish the earth*** *[N3]*, ***and subdue it*** *[N4]*: ***and have dominion over the fish of the sea*** *[N5]*, ***and over the fowl of the air*** *[*N6], ***and over every living thing that moveth upon the earth*** *[N7]*.
>
> (N – Likenesses and H – Living water interactions)
>
> **Genesis 2:11-12 (KJV):** ***The name of the first is Pison*** *[H 1]*: ***that is it which compasseth the whole land of Havilah*** *[H2]...* ***the name of the second river is Gihon*** *[H3]*: ***the same is it that compasseth the whole land of Ethiopia*** *[H4]*. ***And the name of the third river is Hiddekel*** *[H5]*: ***that is it which goeth toward the east of Assyria*** *[H6]*. ***And the fourth river is Euphrates*** *[H7]*.

The rivers of Eden passage above provides a parallel to how Man's living water is formed and it also has transactions and participants; the participants are the *'whole land'*, and the sources of 'good gold', *'bdellium'* and *'onyx stone'*. The *'whole land'* is Man's image life and the sources of the 'good gold' (from Christ), *'bdellium'* (from the Holy Spirit) and *'onyx stone'* (from regenerated Man) make for three more image lives – this makes for participation by four images lives in total ('the four image lives of Pison'). If we strip the 'deliverers' from the 'ten river-water likenesses' we are left with seven river-water likenesses: *'Pison'* (1), *'Havilah (2)'*, *'Gihon' (3)*, *'Ethiopia' (4)*, *'Hiddekel' (5)*, *'Assyria'* (6) and *'Euphrates'(7)*. Because the last three have ambiguity these seven match exactly with the seven likenesses of Genesis 1:28 (at the same time by this ambiguity, *'Euphrates'* can just refer to the Holy Spirit alone or category four of 'one true pattern'). In a way, the river-water interactions are precursors to the seven actual likenesses.

These seven deliverer-stripped water likenesses can be divided into three sets of two followed by *'Euphrates'* on its own: Christ-like traits going out of born-again Man (*'Pison'* + *'Havilah'*); Christ and the Holy Spirit going into Man and regenerating him (*'Gihon'* + *'Ethiopia'*); and the Holy Spirit centred (*'Hiddekel'* + *'Assyria'*). If I use translations of the names of the Eden rivers then, in effect: an 'increase' produces *'Havilah'*; a 'bursting forth' produces *'Ethiopia'*; and 'rapid waters' produce *'east of Assyria'*. The *'Gihon'* and *'Ethiopia'* categories can apply to the Holy Spirit too (as 'active' *'bdellium'*).

Please note again how the *'gold'*, *'bdellium'* and *'onyx'* are just there in the *'Pison'* and *'Havilah'* category; a 'presence' and they may or maybe not all be a part of those interactions. Please note that the use of the word *'the same'* attached to *'Gihon'* may apply to *'Gihon'* itself, may refer to the previously mentioned or *'same'* *'gold'*, *'bdellium'* and *'onyx'*, or may even mean the *'same'* transactions as with *'Pison'*, and in this case they do the *'encompasseth[ing]'* – a more active role. This is all consistent with *'gold'*, *'bdellium'* and *'onyx'* being the 'deliverers' or 'influencers' delivering spiritual gifts contained in their individual selves and moving them in and out of Man, thus onyx (the regenerated Man) may be carrying indwelling gifts from U, C, A or G.

(Please also note that because *'Hiddekel **goeth**'*, it is in a way, its own 'vehicle'.) We thus have another biblically aligned pattern: *'gold'*, *'bdellium'*, *'onyx'* and *'Hiddekel'* (A, G, U ) along with the standard 'one true pattern' (U, A, C and G); and pattern in their paired bases too.

It is worth re-noting a count of these 'one-true-pattern'-aligned interactions ('H's) in the rivers of Eden passage: Category 1 (the first river) has five living water likenesses (two living water interactions and three ‘deliverers’; category 2 (the second river) has two or five interactions depending on if the ‘deliverers’ are counted; category 3 (the third river) has two interactions and category 4 (the fourth river) has one; for a total of ten. I suggest that if an amino acid’s etymology demands the involvement in one or more of these categories then the hydrogen count in that amino-acid side chain will match with the **combined** water count in these categories, likewise for image life and a carbon count, lights and an oxygen count and nitrogen and a likeness count. (All Genesis ‘sets of six’ passages, there are 11, help to this end.)

Also, as the etymology of each amino acid selects certain deliverer-stripped interactions in the rivers of Eden passage, then that number of interactions will correspond **exactly** to the number of codons that code for it. For example, alanine (to be a rock like Christ) requires all four Christ-centred interactions and is coded by four codons.

Each codon’s three bases will also tell an ordered story of saved Man achieving the status or acquiring the gifts that the etymology of the amino acid describes. This makes for several simultaneous patterns in mRNA to tRNA pairing (Figure 2.2 shows 3 paired nucleotides in DNA and this is similar to the situation in mRNA to tRNA pairing but the tRNA additionally carries an amino acid with it.). To summarise these patterns and their directions of flow:

1. Phosphate (Christ), ribose (the Holy Spirit), nitrogenous base (saved Man) – something for the process all men.
2. Phosphate (good gold), ribose (‘*bdellium*’), nitrogenous base (‘*onyx stone*’) – something for Man’s spirit. (This can be looked on as a representing the ‘good gold’, ‘*bdellium*’ and ‘*onyx*’ ‘presence’ in the mRNA nitrogenous base (category 1); and simultaneously, ‘compasseth*[ing]*’ the tRNA base (category 2).
3. Nitrogenous base of mRNA (category 1) ∞ nitrogenous base of tRNA (category 2) ∞ ribose (category 3) ∞ phosphate (category 4) – all making for a ‘one true pattern’.

These run 'along' the paired nucleotides but there is another pattern for each codon – across three bases – and, yet again, the exact same familiar 'one true pattern' appears!

4. All three bases in both mRNA to tRNA pairing will have 'one true pattern' ordering running along their three bases as well as being categories 1 and 2 of 'one true pattern' running 'along' the nucleotide. They will also link to each other in that the category 1 and 3 of the tRNA will be a source of category 1 and 3 of the mRNA; and category 2 of mRNA will be a source of category 2 of tRNA. The mRNA will still be a part of 'good gold', *'bdellium'* and *'onyx'*, and they can be *'encompasseth[ing]'* Man – influencing and changing Man's image life; or in reverse, Man's image life (tRNA) will still be influencing his own 'exported' likenesses, a simultaneous dynamic. Category 3 will represent the Holy Spirit and reveal what He works with, His transactions that may be with any of the five actors in salvation including Himself – and all of these transactions are represented by the third base codon-to-anticodon coupling. (The whole next codon in the mRNA sequence, by its etymology, can then be considered the 'fruitful' category 4 of the 'one true pattern'.)

'Moving' along the nitrogenous bases of the codons will reveal the salvation process in 'one true pattern' ordering. Thus, to manifest what the amino acid etymology demands, the first base of both tRNA and mRNA has to be what comes out of post-salvation Man (category 1) to achieve it, likenesses but still image life. The second base of tRNA and mRNA must be what goes into indwelt Man or is already in Man (category 2), again to achieve the etymological meaning. The third base will represent the required work of the Holy Spirit and will again represent what comes out of Man just like *'Hiddekel'* goeth to *'Assyria'*. Then, with an understanding of defined roles as revealed in the Bible and in the KRM, **these three bases can be predicted** (and maybe even the next codon in the mRNA sequence and position 37 for tRNA too).

**Shed light on Man** (for reference in the amino acid 'stories' below)

1. **A to U:** Christ's (image, likeness, waters or light – summarised as light from here on for these bulleted points), 'the greater light' goes into Man or rules over Man ('the lesser light'); or Man and Christ together. For the whole process it can also be interpreted as Christ and saved Man together.
2. **U to A:** the regenerated Man's light is in, or goes into the 'body of Christ' or is 'a bride for Christ' and becomes more Christ like; or Man and Christ together. For the whole process it can also be interpreted as Christ and saved Man together.
3. **C to G:** the light of the Holy Spirit carrying the Bible reveals the Holy Spirit's light to Man and grows this light in Man; the Bible and the Holy Spirit working together (all men or the individual).
4. **G to C:** the Holy Spirit grows the light of the Bible in Man and interprets it for Man; the Bible and the Holy Spirit work together (all men or the individual).

**Other apparent structural rules**

'$CH_2$' or '$CH_3$' groups are present in amino acids and are called methyl groups and methyl has an etymology of 'wine' and 'wood' (the 'wine from the wood' – a representative reference to Christ's crucifixion). For interactions only (no 'deliverers'), *'Pison'* and *'Havilah'* working on Man's regenerated spirit; *'Gihon'* and *'Ethiopia'* working on Man's regenerated spirit; and *'Hiddekel'* and *'Assyria'* working on Man's whole spirit can all be considered as $CH_2$ events. And the 'deliverers' only can also be considered $CH_3$ events as 'good gold', *'bdellium'* and *'onyx'*, working with any image life also fits as a 'wine and wood' event.

As with DNA, the 'final position' or completion of a certain aspect of salvation is represented by a ring compound, six-ring structures for Man and five for God (or perhaps, God in man), and Jesus Christ referred to this end position as being *'born of the spirit'* in John 3:3-5, (this is consistent with Genesis 1:27, a male body and a female bride).

**More Proof: Individual Tell-tale Amino Acids** (amino acids side chains tell the story of saved Man achieving the status of, or manifesting, what the etymology of the amino acid name describes)

**Either:** Christ and regenerated Man Centred

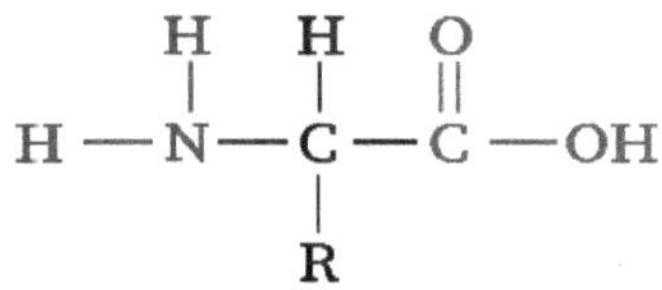

**Or,** Holy Spirit and whole spirit of Man Centred

**Figure 13.2: Amino acid core structure: ProDesigner/Alamy Stock Photo**

Put simply, in order to be a chiral compound, you need a carbon atom to be bonded to four different groups. Amino acids have the same basic core structure. The '**R**' or '**side chain**' varies to form different amino acids and provides a fourth component to achieve chirality. (Think of chirality like our hands: both are hands, but one is left-handed and the other is right-handed – the same but different – the carbon atom (the one attached to the 'R' in figure 13.2) will have four different groups if 'R' is different to the other groups; it will then have a possible mirror-image amino acid. The two mirrored options are called 'enantiomers'.)

In the KRM, RNA and protein synthesis represent working on Man's spirit in the long term – working on the born-again Man – on his regenerated spirit. The core structure in figure 13.2 has a chemical formula that can represent Man's regenerated spirit ($C_2H_4N_1O_2$). The regenerated Man can be considered as being comprised of: his 'regenerated spirit' and this is in *'the body of Christ'* ($C_2$); 'Man's four waters to flow' ($H_4$); his *'every living thing that moveth'* ($N_1$ – Man's dominion over spiritual growth); and the *'greater and lesser lights'* ($O_2$).

But the core structure can also be considered from a Holy Spirit perspective by representing Man's whole or original spirit after salvation only (so only 'Man's four waters to flow' and not six); and in this case the lights change to the Bible and the Holy Spirit, and the image life changes to his whole or original spirit (this can include his regenerated spirit) and a *'bride'*. The centring is not exclusive as the Holy Spirit is represented by *'bdellium'* in all of the Christ centred categories and the Holy Spirit makes saved men a *'bride'* and Christ-like in His own categories too.

When amino acid formulae are divided by viewing this core structure separately, they seem to adhere to very specific times and events of 'one true pattern'. The separated core structure is one aspect of born-again Man; the 'R' or side chain tells his story, and that story will tell you how Man obtains the Christ-like characteristics that the etymological meaning of the amino acid name suggests.

**Image to likeness**

All seven living water interactions and the three 'deliverers and influencers' are all living water likenesses – 'H10' (the living waters made to flow are also 'H's). But, Genesis 1:21 uses *'nephesh chayyah'* in describing category 2 input into Man (*'nephesh chayyah'* translated as *'living creature that moves'* is always a reference to image life). The *'whole land'* of Genesis 2:11-13 is also image life and this means that *'Gihon'* and *'Ethiopia'* can be considered as image-life-based interactions too. This makes for a distinction between the 'interactions' and the 'deliverers' and, in a way, it is an image-to-likeness distinction; they are both living water likenesses but can represent working with image life or likenesses, respectively. Genesis 2:19 confirms a further link in that likenesses are made from image life (*'ground'*).

Hence, a $CH_2$ referring to *'Gihon'* and *'Ethiopia'* or *'Pison'* and *'Havilah'* are image statements. A $CH_3$ may be referring to image life making the three likenesses of 'good gold', *'bdellium'* and *'onyx'*. A 'H–C–$CH_3$' may be considered as two image lives and four 'image' water interactions – an image statement.

Genesis 2:19 ('beast' and 'fowl') describes two likenesses being made from image life (ground), a $CH_2$ could be describing this process.

{**Legend: for codons: A** - Christ as God ∞ **C** - the Holy Spirit with the Bible ∞ **G** - the Holy Spirit ∞ **T** - Man or Jesus as a man ∞ **U** - regenerated Man

**For chemical formula: C** - image life; **H** - water likenesses; **N** - likenesses; **O** - lights

**Codon ↔ anticodon** is category 1 ↔ category 2 (a dynamic of one true pattern)

**Codons and anticodons:** First base – what comes out of man ∞ second base - what goes into the now indwelt Man ∞ third base – the working Holy Spirit (a one true pattern). Codons are biblically-defined internal interactions.

**Etymology:** The amino acid etymology will select certain rivers and certain essential actors going in and out of indwelt Man

**Chemical formula predictions**: Chemical formula will be deducible from the allocations of who does what within the biblical 'one true pattern' statements; coupled with knowledge of what building blocks are needed to manifest the etymological meaning of the amino acid, internally, externally or both.}

Please note that: Man's body can make 'non-essential' amino acids: alanine, asparagine, glutamic acid and aspartic acid (and sometimes serine is included here). Man's body cannot always make 'conditionally essential' amino acids: arginine, cysteine, glycine, glutamine, tyrosine, proline and serine. Man's body cannot make essential amino acids, and they have to be ingested: valine, isoleucine, leucine, methionine, phenylalanine, threonine, tryptophan, histidine and lysine.

***Amino acid derives its name from 'amine', which has a history associated with the personal name 'amine' and also 'ammonia'; it has meanings of 'trustworthy', 'hidden one', 'support', 'creation' and 'faithful one';*** I suggest 'hidden one' is the most pertinent as Christians are referred to as 'hidden' in the '*body of Christ*' in Colossians 3:3.

**A profound aside:** Please note how in Genesis 2 (KJV) the only times the word 'became' is used are: '***man became a living soul***' and '***river became into four heads***'. These two are the most profound moments of the whole salvation process. Please also note that the only two common amino acids that contain sulphur ('S' – the washed-clean Man or Man manifesting a washed-clean status) are methionine and cysteine, which, by their etymology, tell the same story as the two Genesis 2 quotes, respectively.

**Methionine** (born-again and washed-clean Man)

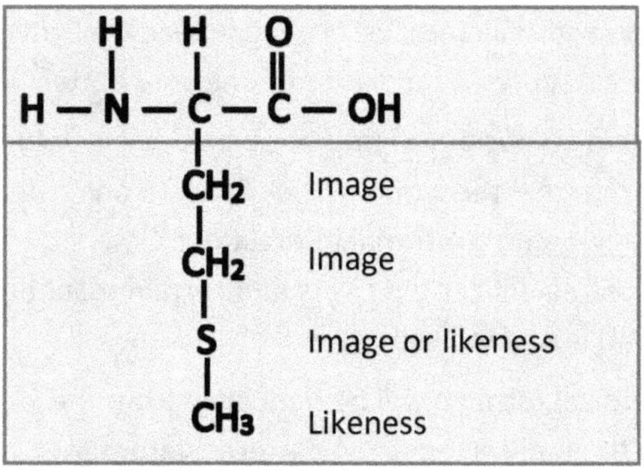


**Figure 13.13: Methionine**

Methionine is coded by AUG (in mRNA), and, as previously discussed, this pairs with UAC (in tRNA) – the 'birthing sequence'. All protein synthesis (metaphorising 'of Christ') begins with methionine: a trade-mark structure – 'one true pattern'.

Methionine confirms the pattern that will lead to predictability of the nitrogenous bases in each codon. As previous, the anticodon UAC shows 'one true pattern': what comes out of Man (U); what goes into Man or is in Man (A); the Holy Spirit's work (C). I suggest this pattern happens in all codons and anticodons, for methionine the codon AUG will then exhibit the same three dynamics. 'A' – Christ-like likenesses come out of Man, 'U' – Man is placed in Christ (the body or bride depending on context), and C the working Holy Spirit. The amino acid etymology will reveal which interactions the Holy Spirit needs and which participants are required and these can then be ordered appropriately. A changing third base then becomes a way of counting Holy Spirit-led interactions or 'counting codons'.

I am now going to demonstrate that the number of codons that code for each amino acid, the nitrogenous bases within that codon and the chemical formula, and even chemical structure of all amino acids are consistent with the text in Genesis 1 and 2. This will be achieved by 'determining' 'one true patterns' running in different directions and by comparing them with biblical 'one true patterns'.

Methionine is the only amino acid that **has** to be 'read' from both Christ and Holy Spirit perspectives simultaneously. The inclusion of sulphur (S) (Man actually being made in the etymology) necessitates this as both the Holy Spirit and Christ are needed to this end. To reemphasise, in other cases certain amino acids **have** to be read **individually** in two ways, and in some cases amino acids **can** be read in two ways. In these cases a '$CH_2$' may switch from *'Pison'* and *'Havilah'* to *'Hiddekel'* and *'Assyria'* with *'Gihon'* and *'Ethiopia'* remaining in both, respectively (I will term this 'a Holy Spirit switch' from here on). For valine, phenylalanine, threonine and histidine, they all have to be read in two ways (because their etymology applies to both Christ and the Holy Spirit, individually and not simultaneously). Similarly, some other etymologies will select contributions from **both** Christ- and Holy Spirit-centred categories by 'crossing over' and featuring in all six living water interactions (*'Pison'*, *'Havilah'*, *'Hiddekel'*, *'Assyria'* with *'Gihon'* and *'Ethiopia'*). This occurs in lysine, leucine and isoleucine. Please note that all amino acids mentioned by name in this paragraph are essential amino acids. I will also show that categories 2a- and 2b-assigned amino acids are those that **can** be read in two ways.

**The etymology and codon count of methionine** specifically points to a Man washed clean ('thios') by the blood of Christ ('methyl'), and this aligns, also specifically, with Ethiopia (described in Genesis 2:7 as *'man became a living soul'* and in Genesis 1:28 as *'subdue'[d]*). Methionine as one entity itself (the washed clean Man) is coded by just one codon but as well as this '2b' (Ethiopia), methionine needs the previous three Christ-centred interactions to 'get there'.

**Side chain element count:** the etymology demands that all Christ centred water likenesses are required (H1-H4) to 'get there' and also the 'three vehicles' – $H_7$. The 'four image lives of Pison' are all used in this process but the change in Man's status is recognised by a change to a washed clean Man and Man's original spirit life (C) becomes an 'S', hence, $C_3H_7S_1$ is methionine's side chain.

(The element count will hold for the Holy Spirit perspective too by considering the Holy Spirit's role in 2a and 2b.)

**The chemical structure story** (see the above image): Because elements are generic, they can metaphorise in a second, and chronologically correct, 'story'. The born-again Man (amino acid core) has two living water interactions (*'Pison'* and *'Havilah'*– $CH_2$), has two more interactions (*'Gihon'* and *'Ethiopia'*– $CH_2$) and is washed cleaned (S) by these and displays this 'cleanliness' through his regenerated spirit (the *'out of the ground'*) of Genesis 2:19 through three 'vehicles' ($CH_3$). Likenesses begin after the last carbon (called a delta or 4$^{th}$ carbon) atom. (I suggest that sulphur (S), like 'H', can be an image or a likeness.) Please also note that the amino acid diagrams are just two-dimensional images of three-dimensional compounds, and perhaps it is best to consider the likenesses ('H's) as on the 'outside'.

**Alanine** (born-again Man, an evangelising 'little rock' in the church)

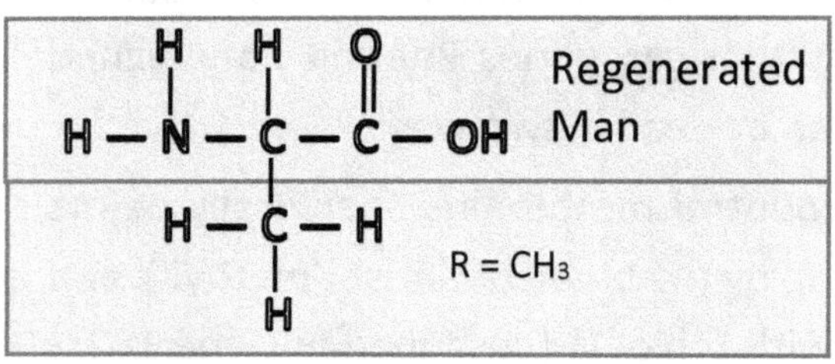


**Figure 13.3: Alanine**

Alanine aligns with category 1a.

**The etymology:** to be a 'rock' (a 'rock' in the church) requires Man to have all four Christ-centred interactions (H1 to H4), but the church aspect of the etymology also means an element of evangelisation. The Bible describes Peter and Christ as *'this rock'* (Matthew 16:18); and the Bible tell us[71] that only the Holy Spirit can save Men and evangelise (men only help and 'fruitful' is evangelising); and also that the Bible itself is needed for faith[72], G and C must feature in the codons and anticodons.

---

[71] 1 Corinthians 12:3: ***Wherefore I give you to understand, that no man speaking by the Spirit of God calleth Jesus accursed: and that no man can say that Jesus is the Lord, but by the Holy Ghost.***

[72] Romans 10:17: ***So then faith cometh by hearing and hearing by the word of God.***

**Codon count:** alanine (to be a rock) demands the four Christ-centred living water interactions (H1 to H4) – alanine is coded by four codons.

**Side chain element count:** the etymology demands evangelisation (*'fruitful'* (N1) in Genesis 1:28), the matching by category living water likeness is H1 (*'Pison'*), and this is described in Genesis 2:11-12 as being made from Man's spirit (whole land) – $C_1$, in the presence of the three water vehicles $H_3$. The matching category passage in Genesis 2:7 also confirms *'Pison'* is made from 'dust of the ground' (Man's spirit). The side chain of alanine is, predictably, $CH_3$.

**Nitrogenous base reconciliation:** to be a 'fruitful' Man, Man must have the Holy Spirit working through him, this means the Holy Spirit (G) must be the main player in likenesses coming out of Man or category 1 (or the codon); and the Holy Spirit works with the Word of God (C) to this end. The first nitrogenous base must be 'G' and the second must be 'C' in that order in the codon. Alanine is coded by GC (U, C, A or G) in consistency with this. The third base represents the Holy Spirit's work in evangelising represented by the four interactions in each codon and with their complementary pairing. These can be matched with the specific interactions in the rivers of Eden passage both quantitatively and qualitatively. All of these interactions together will make Man an evangelising 'little rock' in the church. The match with the rivers is: U to A aligns with *'Pison'* (an increase and fruitful); C and G (gluing) aligns with *'Gihon'* (bursting forth and replenish); A to U aligns with *'Havilah'* (together and multiply); and G and C (gluing) aligns with Ethiopia (Ethiopia and subdue); please see the previous page 'shed light on Man' for details of interaction dynamics.

(I will term these interactions 'the Holy Spirit selected living water interactions – one'. Similarly, some Holy Spirit aligned interactions may incorporate *'Hiddekel'* and *'Assyria'* instead of *'Pison'* and *'Havilah'*, in this case the body of Christ is replaced by a bride for Christ. I will term this second batch of interactions 'the Holy Spirit selected living water interactions – two'.)

**The chemical structure story:** as a part of 'the overall process', the born-again Man (amino acid core) gets an increase ($CH_3$) via the Holy Spirit to become an evangelising little rock in the church (alanine). The central core is the image (C) and the periphery ($H_3$) is the likeness, and image-to-likeness structure is satisfied.

## Glycine (born-again Man made 'sweet' by the Holy Spirit)

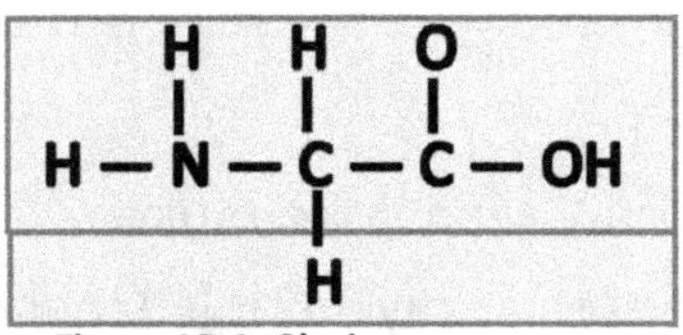

Figure 13.4: Glycine

**Chirality**: for glycine there is no L-chirality; the representations in the core structure now change to being centred on the Holy Spirit.

This is the Holy Spirit at work and the 'sweetness' is now not from Christ and this is confirmed by glycine's achirality. This will be the first switch in the representation of the core amino acid structure from Christ- and *'body-of-Christ'* centred to Holy Spirit- and 'bride-for-Christ' centred; it will also occur with arginine, cysteine, glutamine, proline, serine and tyrosine – and I will show that this is significant because these are all of the 'conditionally essential' amino acids – and all representatively, will have a Holy Spirit theme (I ask you to consider the use of *'dominion'* in Genesis 1:28 too).

**Etymology:** To be 'sweet in the eyes of God' ('glycine'), Man needs to have God-like traits (Ephesians 5:2). In a way, Man receives 'light' in all 'rivers', but specifically the Holy Spirit gives something – making Man 'sweet' in the eyes of God, in *'Hiddekel'* and *'Assyria'* (H5 and H6), and also his role as the glue in *'Gihon'* and *'Ethiopia'* (H3 and H4). The Holy Spirit's role is more of a presence and not an image-life gift in *'Pison'* and *'Havilah'*.

**Codon count:** the etymology of glycine demands the four proactively-given, and Holy Spirit-centred living water interactions (H3 to H6) and glycine is coded by four codons.

**Side chain element count:** in terms of living water the Holy Spirit gives living-water-making capabilities to man (H). The side chain is just one 'H' (this also gives achirality). The side chain is just $H_1$ (I should point out that normally if there is just one atomic element then the number is omitted but I will continue to include it in chemical formula to emphasise the counting process).

**Nitrogenous base reconciliation:** God describes all things to do with salvation in Man as 'sweet smelling' (Ephesians 5:2 and 2 Corinthians 2:15) and the achirality rules out anything of Christ. This demands that the Holy Spirit and Christ are present in Man (C in category 2 or the anticodon) and this will pair with 'G' in the codon (the Holy Spirit gives to Man). The Holy Spirit Himself carries out this process and so this confirms that the two codon bases must both be 'GG', the Holy Spirit going in and out of Man (as opposed to proline that has both the Holy Spirit and the Bible coming in and out of Man). Glycine is coded by GG (U, C, A or G) in consistency with this. The 'the Holy Spirit selected living water interactions – two' apply.

**The chemical structure story:** As a part of the overall process, born-again Man (amino acid core) by the Holy Spirit will be given indwelling Holy Spirit-like gifts and biblical living-water-making gifts (H) 'sweet smelling' to God (glycine). 'H' can be either an image or likeness in itself and hence, can exhibit image-to-likeness structure depending on how it is interpreted for each water-making gift.

## Valine (born-again Man gets full spiritual health)

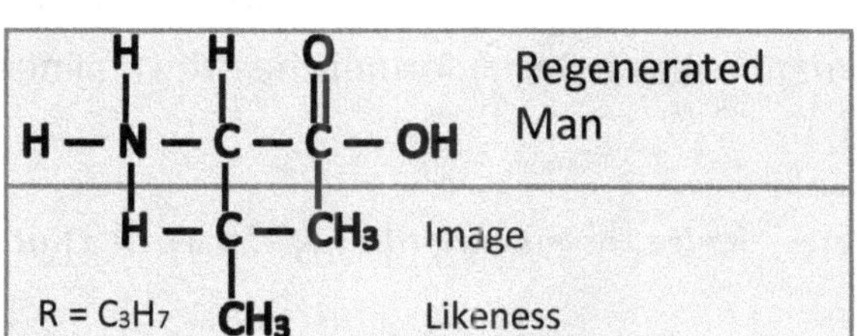


**Figure 13.5: valine**

**Etymology:** To become spiritually 'healthy' must be considered and read as both Christ and Holy Spirit centred. The Bible describes theitself[73] and the Holy Spirit [74] as pivotal for spiritual growth – there must be a 'C' in Man (anticodon). To be healthy requires Christ in Man and Man in Christ (body and bride) – there must also be an 'A' in Man.

**Codon count:** Christ-centred interactions are (H1 to H4); and Holy Spirit centring are (H3-H6); and valine is coded by four codons.

**Side chain element count:** The Christ-centred categories have seven likenesses in total, and for the Holy Spirit centred count, H5 and H6 replace H1 and H2 to maintain this count – $H_7$. For spiritual health Man uses all 'four image lives of Pison' but valine splits these into Christ and Holy Spirit centred, the former has no Holy Spirit image life and the latter no Christ image life – $C_3$ for both. Valine's side chain is $C_3H_7$.

**Nitrogenous base reconciliation:** mRNA codons are GU with (U, C, A or G). The etymology demands that the Holy Spirit, the Bible and Christ are in Man (category 2) and so CA must be in the anticodon. This is consistent with GU (in the codon), the Holy Spirit manifesting from Man and Man being in the body of Christ or the bride of Christ depending on centring. Both 'the Holy Spirit selected living water interactions – one and two' apply, as appropriate.

**The chemical structure story:** Born-again Man (amino acid core) gets acted on by Christ's and the Holy Spirit's image life ($C_2$) and has four interactions or $H_4$ (*'Pison'*, *'Havilah'*, *'Gihon'* and *'Ethiopia'*, and the first two exchange with *'Hiddekel'* and *'Assyria'* for Holy Spirit interpretation) and these are represented by a '$H–C–CH_3$'. Man, by his regenerated spirit or $C_1$ (the image ends here and the likeness begins) has three 'vehicles' ($H_3$) to manifest as being spiritually healthy (valine).

---

[73] 2 Peter 2:2: ***As newborn babies, desire the sincere milk of the word, that ye may grow thereby.***

[74] 2 Corinthians 3:18: ***But we are all, with open face beholding as in a glass the glory of the Lord, are changed in to the same image form glory to glory. Even as by the spirit of the Lord.***

**Aspartic acid** (*'2b'*) **and glutamic acid** ('2a'); – saved Man acquires the *'lesser light'* and the *'greater light'*

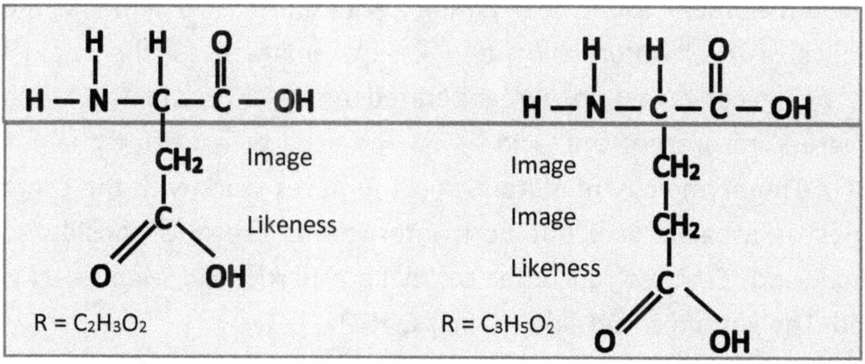


**Figure 13.6: Aspartic acid and glutamic acid**

Both of these 'R' or side chain components deprotonate and lose one hydrogen ion to form aspartate and glutamate. The amino acids can represent: 'the born-again Man as a 'light bringer' by the regeneration of Christ's light on his own spirit (aspartic acid – **a '2b'**) or the image life of Christ as a light bringer in Man and becoming glued to Man; Man within Christ and in the *'body of Christ'* (glutamic acid – **a '2a'**). All give 'light' ($H^+$). Deprotonation will reveal forward progress in salvation.

**Etymology:** the use of 'acid' means a 'dealer' in the 'currency of light' and so both amino acids must contain oxygen (O). Genesis Day 4 describes God making a *'greater light'* and a *'lesser light'* with the two having a shared function in that the greater rules over the lesser. Both of these amino acids must have a second 'O' to incorporate this dynamic. The 'aspartic' is a reference to saved Man as a growing shoot becoming an 'acid' (1 Corinthians 3:6-9) and the 'glut' references Man's spirit being glued to Christ.[75] It is the Holy Spirit that does the gluing and yields the acid, but it is Christ that is the entity that Man is glued to, and so 'G' is what comes out of Man to give light (codon) and 'A' is what goes into Man (anticodon as the light source), respectively.

---

[75] 1 Corinthians 6:17: ***But he that is joined unto the Lord is one spirit.***

**Codon count:** the etymology of both demands the use of just the two category 2 living-water interactions or image-life interactions (*'Gihon'* and *'Ethiopia'*), and both amino acids are coded by two codons.

**Side chain element count:** the etymology of aspartic acid demands the presence of the *'two great lights'* – $O_2$; two image lives (the original spirit being converted to a regenerated one) – $C_2$; and the three 'deliverers' to present the 'acid' – $H_3$. The aspartic acid side chain is $C_2H_3O_2$. The etymology of glutamic acid requires work with the same entities as aspartic acid but both interactions are also considered (*'Gihon'* and *'Ethiopia'*) in order to 'make aspartic-acid Man' – $H_2$ is added. The glutamic acid side chain is $C_2H_5O_2$.

**Nitrogenous base reconciliation:** the mRNA codon is GA in the first two bases for both with (U and C the third base for aspartic acid and A and G for glutamic). The Holy Spirit works with Christ to make Man born again and release 'light' either into Man for glutamic acid or into the world for aspartic, the first base must be 'G'. The second base moves the other way (into Man) and must be Christ going into Man (A to U) for both amino acids. The third base reveals the dynamic (again going out of Man) and note the way it is now Christ (A) and the Holy Spirit (G) coming out of Man to be joined with Man for glutamic acid, and it is the regenerated Man (U, carrying Christ like gifts) and C (the Holy Spirit and the Bible) manifesting 'out of' saved Man.

**The chemical structure story:** For aspartic acid, the born-again Man's original spirit ($C_1$), by the work of *'Gihon'* and *'Ethiopia'* ($H_2$) which regenerate Man's spirit (another $C_1$). The image ends there and likenesses begin: Man manifests the greater and lesser light ($O_2$) and as a living water vehicle (H), the onyx stone. For glutamic acid an extra image based $CH_2$ is added to represent the Holy Spirit's perspective in His gluing action in order to make 'aspartic acid'. Please note that, as the Bible and the Holy Spirit give light to Man, then aspartic acid and glutamic acid **can** also be read from a Holy Spirit perspective, the two lights changing to the Holy Spirit and the Bible.

**Asparagine** (*'Havilah'*) **and glutamine** (born-again Man as a 'shoot', grows and made a 'hidden part' glued to God)

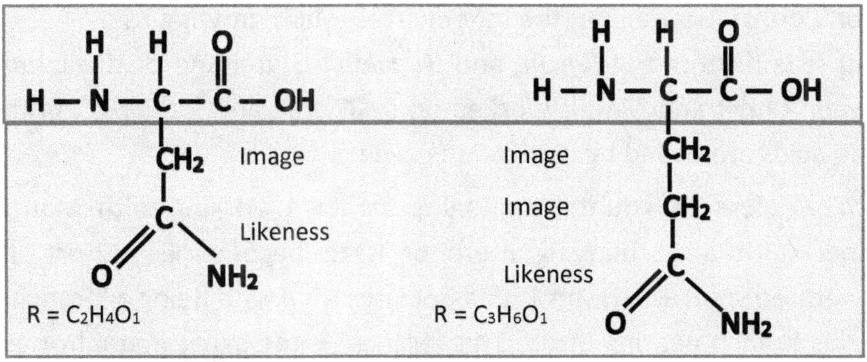


**Figure 13.7: Asparagine and glutamine**

Asparagine will be shown to match with '1b' and is considered non-essential. Please notice the similarities and differences between asparagine and glutamine compared with aspartic acid and glutamic acid. The latter are dealing with both the input of light and the making of 'lights'; the former are just dealing the manifestation of these lights. This is represented in just $O_1$ to reflect asparagine and glutamine working on one light (the lesser light and the greater light, respectively). In both cases the hydroxyl group (OH) is replaced by an amino group ($NH_2$). Man has to use his own likeness ('N') to manifest this 'end' described in the etymology of the amino acid, a likeness under Man's volition. Please note the way *'kol'* (which represents an interaction with Man) is the root word *'whole'* in Genesis 2:11-13.

**Etymology:** I suggest the etymology of glutamine leads to a Holy Spirit-centred interpretation because the 'gluing' itself is a Holy Spirit activity, albeit 'gluing' Christ to man in this case. I have discussed above how the 'glue' applies to: 'gluing' Man's spirit to 'Christ's image life', gluing or joining Christ to the regenerated Man and 'gluing' Man in the *'body of Christ'* (glutamine is similar to oxalosuccinate in Krebs cycle). This means that in a Holy Spirit-centred interpretation; the amino acid core can switch to the whole spirit, but will still represent hydrogen counts associated with living water gifts ('H' counts) given by Christ centring.

Asparagine – working on the 'shoot', and glutamine – working on the 'bond'. Glutamine is considered 'sometimes' essential.

**Codon Count:** Asparagine (the regenerated Man) growing as a 'shoot' in Christ is nurtured in *'Pison'* and *'Havilah'*. Glutamine or the bond between Christ and Man is worked on in *'Gihon'* and *'Ethiopia'*, both amino acids are coded by two codons each.

**Chemical element count:** Asparagine involves working with Man's original spirit and changing more of it to become like Christ, a regenerated spirit – $C_2$. John 15:5 associates Man with being a 'branch' and this branch bearing 'fruit'. This 'branch' is the same metaphor as 'asparagine' and hence, asparagine must multiply to bear fruit and have an 'N' (N2 – multiply) – $N_1$. Asparagine is in category 1 and hence, must also have $H_2$ to contain the two needed interactions of *'Pison'* and *'Havilah'* – $H_2$. Genesis 1:18 tells us that a 'light' is needed to rule over the darkness, the lesser light – $O_1$. The, now, 'shoot Man' (and not the 'onyx man' and the normal two other 'vehicles' of a '$H_3$') has a 'shoot' as one of the vehicles and $H_3$ becomes – $NH_2$. The side chain of asparagine is $C_2H_4O_1$. For glutamine, there are two '$H_2$'s, one each for Christ and the Holy Spirit, both with representations in *'Gihon'* and *'Ethiopia'* – $H_4$. Three image lives are placed or worked on in Man, Man's original spirit, his regenerated spirit and Christ's image life. The final group (C, O and $NH_2$ group) differs from asparagine in that the 'N' changes to represent *'subdue'* and the 'O' to Christ (the 'greater light').

**Nitrogenous base reconciliation:** The essential element of both etymologies is growing in Christ or being one (or glued) to Christ for asparagine and glutamine, respectively. Christ must be revealed in Man in **asparagine** A from U (anticodon to codon) as the main actor. But Christ is also the mechanism for this action and must again feature in the second base A and U again (Christ into Man). The third base interaction must match with *'Pison'* and *'Havilah'* as these are the two counting codons, and please note how AAU with the third base U to A pairing is consistent with Man being placed in the body of Christ (this is consistent with Man's likenesses being 'beast of the earth' as previously found). The change in the third base for the second codon is also consistent, as found with alanine, the Holy Spirit and the Bible using Man to be 'fruitful'.

For glutamine, represents the Holy Spirit and Christ in the gluing of Christ and Man together, so both must feature in the codon or anticodon. The overriding summary of *'Gihon'* is, by its etymology, life bursting forth into Man and in abundance too. These 'living waters' that carry out this transportation are the Bible and Holy Spirit as described in Genesis Day 5 and in Ephesians 5:26, hence, they are the main actors of the 'gluing' (not the indwelling gift, that is Christ). 'C' must be the first letter of the codon and 'A' (Christ going into Man) must be the second. The third base interaction again shows the Christ centred (A to U) to represent *'Gihon'*, and the Holy Spirit centred (G to C) to represent the actors in regenerating Mans spirit in *'Ethiopia'*.

**The chemical structure story:** the born-again Man (amino acid core) has two interactions and has Christ image life glued to his own and becomes a growing fruitful branch in the 'body of Christ'. Image to likeness changes take place in the final group with 'O' and '$NH_2$' as likenesses.

## Serine and cysteine (Man as a *'bride'*, and a living-water vessel)

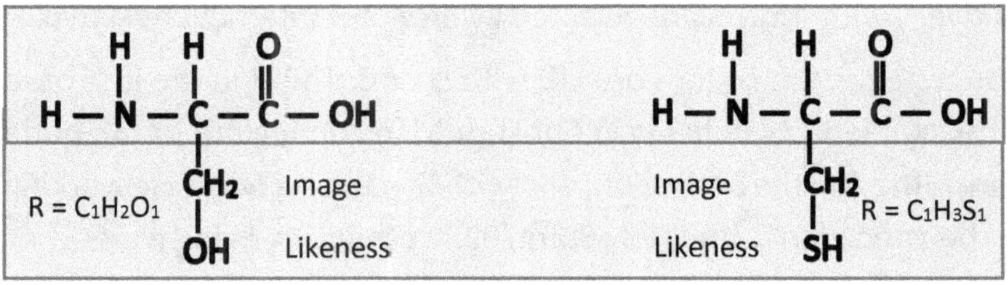


**Figure 13.10: Serine and cysteine**

**The etymology:** Serine and cysteine are by etymology (adorning a bride and a 'carrier of water'), and both of these are Holy Spirit centred (from the parable of the ten virgins and John 7:37-39, respectively). As well as this, in order to be a bride Man must have all four Christ-centred interactions (Ephesians 5:22-33 describes this 'oneness' with Christ). The appearance of sulphur (S) in the chemical formula confirms Man has to be washed clean in order for the waters to flow.

**Codon Count:** Serine (to become and adorn a bride) demands all four Christ-centred living water interactions and both Holy Spirit centred ones – serine is coded by six codons (H1-H6). Cysteine just requires the two Holy Spirit living water interactions (also confirmed in *'let the waters bring forth'* of Day 5) and is coded by two codons (H5 and H6).

**Side chain element count:** Serine is representing what adorns Man as a bride and this involves just the 'deliverers' that reveal it – $H_3$; works on Man's whole spirit – $C_1$; and make Man a 'light full of oil (for light)' from the ten virgins parable – $O_1$. The serine side chain is $C_1H_3O_1$. By the association of 'became' with making living water flow, cysteine's representation contains $S_1$, cysteine needs a washed clean Man and serine adorns a bride. The hydrogen and carbon representations are the same – cysteine side chain is $C_1H_3S_1$.

**Nitrogenous base reconciliation:** The six codons of serine have UC or AG in their first two bases. The UCU, UCC, UCA and UCG are consistent with the regenerated Man (U) and the Holy Spirit and Bible (C) working together in the first two rivers, and 'the Holy Spirit selected living water interactions – one' apply. The AGU and AGC are consistent with the third river', with 'Holy Spirit selected living water interactions – two'.

For cysteine, the codons are UG with U or C. The U in the first base is consistent with Man being in Christ or Man coming out of the bride of Christ (the A in the anticodon) and with G in the second base working with the anticodon C (the Holy Spirit/Bible producing living water).

**The chemical structure story:** The Holy Spirit works on the whole spirit of Man (C) in two specific living water interactions ($H_2$) to make saved Man adorned like a bride and make Man a 'light' and a giver of living water (H), (or a washed clean Man and a giver of living water).

**Proline** (born-again Man gains (or maybe is baptised by) fire and oil by the Holy Spirit, this may be a reference to Luke 3:16 '...***and with fire***')

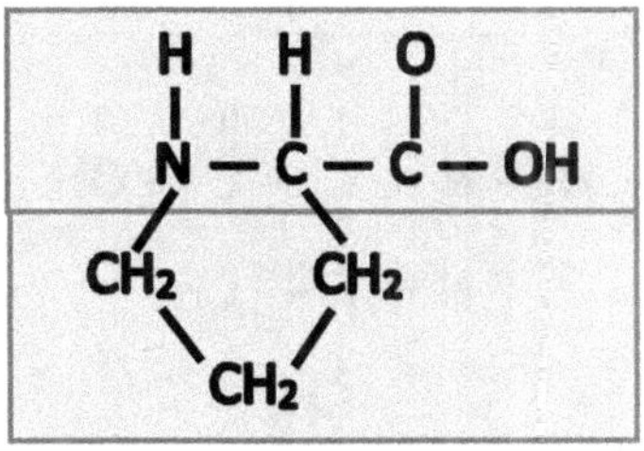


**Figure 13.8: Proline**

**A summary:** 'Fire' is associated with the Holy Spirit (Acts 2:3-4, Matthew 3:11 and 1 Thessalonians 5:19) and 'oil' is too in the Parable of the Ten Virgins. Proline thus, requires that the Holy Spirit alone is in Man and comes out of Man (GG in the first two bases of the anticodon) with the help of the Bible (CC in the codon). The presence of a five-sided ring (a pyrrolidine ring) is consistent with representation of the Holy Spirit as a separate entity dwelling in Man, and a different relationship to Christ who gives his image life to Man and that image life is then Man's. The core amino acid represents four living waters to flow, but the Holy Spirit has five living waters to flow; the looping back of the side chain to the core and a sharing of one 'H' is consistent with this upward change.

The Holy Spirit proactively acts in H3 to H6 and proline is coded by four codons. The Holy Spirit is 'gluing' to Christ in some of these interactions and 'the Holy Spirit selected living water interactions – two' apply. So, to amplify these interactions a little, for this third base interaction and from a 'codon' viewpoint: U to A, Man glued to Christ and is in a 'body' ('*Ethiopia*'); C to G, the Bible and the Holy Spirit as the vehicles revealing the Holy Spirit ('*Hiddekel*'); A to U, Christ is put in Man ('*Gihon*'); and G to C, the Holy Spirit giving gifts of the spirit and biblical knowledge ('*Assyria*').

The Holy Spirit gives his own image life, works with Man's whole spirit and uses Christ's image life – $C_3$; the four living water interactions in the second and third river apply (H3-H6) and the only vehicle is '*bdellium*' – $H_5$. The born again Man is sealed to Christ by the Holy Spirit ($CH_2$), has various Gifts for the Holy Spirit ($CH_2$) and the Holy Spirit lives in Man and this presence baptises Man with fire and is likened to oil for the lighting of a lamp in the Bible. There is no specific side chain.

**Leucine** and **isoleucine** (born-again Man is as white as snow)

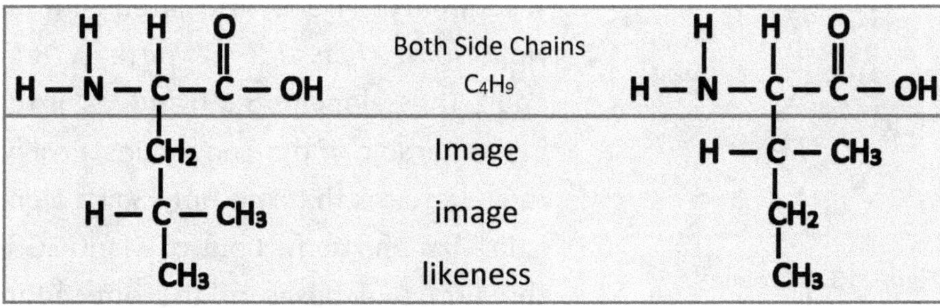


**Figure 13.12: Leucine and isoleucine**

Please note the use of 'the same' in Genesis 2:13, and the use of 'iso' (meaning 'the same') here and its use with citric acid in the Krebs cycle, thus, in consistency, leucine is light in Man and isoleucine is the 'same', but as a sub-category of what goes into Man to achieve it. As previous again, the washing clean occurs in the first three rivers – regenerated Man being made more Christ-like and 'as white as snow' (Daniel 12:10, Isaiah 1:18 and Revelation 1:18).

**Summary:** to become as white as snow involves the work of Christ and the Holy Spirit and this is described in all of the six living-water interactions that are associated with the first three rivers, leucine is coded by six codons. As a subset of this, isoleucine describes what goes into Man to this white-as-snow end, *'Gihon'*, *'Ethiopia'* and *'Assyria'* all go into Man and isoleucine is coded by just three codons. In these three rivers there are nine river-water likenesses in total and leucine and isoleucine have – $H_9$; all the 'four image lives of Pison' are used – $C_4$, the side chains for both are $C_4H_9$.

For leucine, the first two bases in two of its six codons begin UU, representing 'whiteness' coming from Christ being in Man (*'Pison'*) and Man and Christ together (*'Havilah'*) in the first river. The next four codons all have CU in the first two bases and this is consistent with the

Holy Spirit working, or making white, the regenerated Man by either gluing him to Christ or in the body of Christ (*'Gihon'* and *'Ethiopia'*), or making a bride for Christ and giving gifts of the spirit (*'Hiddekel'* and *'Assyria'*). Isoleucine, changes the direction of flow and in line with 'one true pattern' focuses on only what goes into Man but then on how these entities effect what goes out of man in order to be 'white' in the outside world. The first two bases are AU and reference Christ in Man and Man in both the *'body'* and a *'bride'* for Christ.

For leucine, the born-again Man has two interactions with God ($CH_2$ in the first river) and four more interactions ('H–C–$CH_3$' in the second and third river) and uses the three vehicles to make himself as one who *'shall be as white as snow'* ($CH_3$). For isoleucine, the direction is the other way, the born-again Man has four interactions ('H–C–$CH_3$' in the second river and third rivers) and two further interactions (the first river ($CH_2$) to work in the outside word as apart of God's plan and to present themselves as *'washed their robes and made them white in the blood of the Lamb'* ($CH_3$). (Please note that some of the chemical formula stories may also follow 'one true pattern' categories.)

**Lysine and arginine** (born-again Man gets the 'sin' broken down and refined like silver by the Holy Spirit)

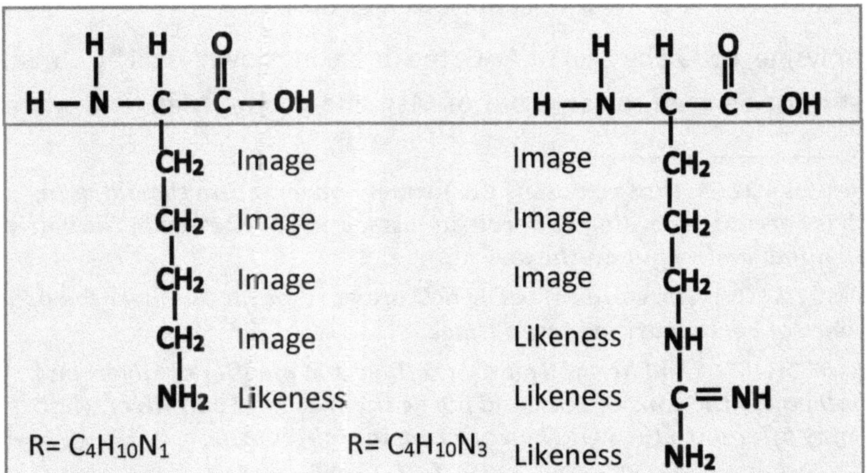


**Figure 13.9: Lysine and arginine**

The Bible describes Man's sinful nature being broken down by the salvation process enabling Man to live 'in the spirit' and lysine has etymology that supports its use as a representation of Man having this trait.[76] The Bible mentions Man's spirit being refined like silver[77,78] and the Holy Spirit as the one who purifies (sanctifies) Man[79], and this is consistent with arginine representing a Man carrying a purified status and that purification is by the Holy Spirit.

**Summary:** Man can be considered as washed clean by the Holy Spirit in all of the six interactions in the first three rivers and arginine is coded by six codons. Man's sinful spirit is transformed by Christ in Ethiopia ('*subdue*') but also by the Holy Spirit in '*Assyria*' ('*fowl of the air*') and lysine is coded by two codons. Lysine is, thus, one of the amino acids that **has** to acknowledge this, with Christ and Holy Spirit centring.

The six codons that code for arginine all have 'G' as second base in their codons and this is predictable in that the Holy Spirit going into Man and working with what is at the first base during this process. Four of the six first bases in the codons are 'C' and two are 'A' and these are reflecting the actions in the codon-representing interactions, Christ centring in '*Pison*' and '*Havilah*' and Holy Spirit centring in '*Gihon*', '*Ethiopia*', '*Hiddekel*' and '*Assyria*'. The 'Holy Spirit selected living water interactions – one or two' will apply, as appropriate.

For lysine, coded by AAA or AAG, the 'breaking down' is all by Christ, Christ must be seen to come out of Man and go into Man, and when

---

[76] Ephesians 4:22-24: ***that ye put off the former conversation the old man, which is corrupt according to deceitful lusts, and be renewed in the spirit of you mind, and ye put on the new man …***

[77] Psalm 12:6 (KJV): ***The words of the LORD are pure words: as silver tried in a furnace of earth, purified seven times.***

[78] Malachi 3:3 (KJV): ***And he shall sit as a refiner and purifier of silver: and he shall purify the sons of Levi, and purge them as gold and silver, that they may offer unto the LORD an offering in righteousness.***

[79] Romans 15:16: … ***that the offering of the Gentiles might be acceptable, being sanctified by the Holy Ghost.***

going into Man it may be into the 'body of Christ' for Christ centring or as a 'bride' for Christ for Holy Spirit centring. 'A' must be at first and second bases of the codons. The Christ-centred relevant codon (*'Ethiopia'* or *'subdue'*) requires more of Man being placed in the body of Christ (U to A) and for Holy Spirit centring (*'Assyria'* and *'fowl of the air'*) requires the Holy Spirit and the bible working together (C and G).

The chemical formula of lysine can be predicted as lysine can represent working through each stage of 'one true pattern' to achieve this removal of the sinful element in Man. Thus, all ten river water likenesses, the four image lives of Pison, and specifically, Man's own likeness requirement to contribute to this end is recognised by the inclusions of one 'N' (*'subdue'*) – $C_4H_{10}N_1$ is the formula of the side chain of lysine. Born-again Man interacts with river 1 ($CH_2$), river 2 (twice, once with Christ and once with The Holy Spirit) (CH2 and CH2), river three ($CH_2$) and attains living in the spirit status manifesting as an 'old self removed' Man ($N_1$) through 'good gold', and *'bdellium'* ($H_2$).

For arginine, the chemical formula also goes through each stage of 'one true pattern' and hence also has all ten river water likenesses and 'the four image lives of Pison' too. But now three more likenesses are needed to represent Man who needs to contribute to this purification (the three likenesses that occur after the use of *'dominion'* in Genesis 1:28, N5, N6 an N7 – *'fish of the sea'*, *'fowl of the air'* and *'every moving thing'*) – $C_4H_{10}N_3$ is the side chain for arginine. Born-again Man gets refined like silver in river 1 ($CH_2$), river 2 ($CH_2$), river three ($CH_2$) and contributes with his own 'old self removed' Man (N) and through 'good gold', and *'bdellium'* ($H_2$) manifests as being purified like silver and is then sanctified by the Holy Spirit.

**Histidine** (born-again Man presents as weaved together with God)

The weaving together to become one is consistent with the events of *'Gihon'* – these rivers describe Jesus Christ's image life and Man's spirit being knitted together, and also Man's spirit being placed in *'the body of Christ'*. But Man can also be considered as being weaved together

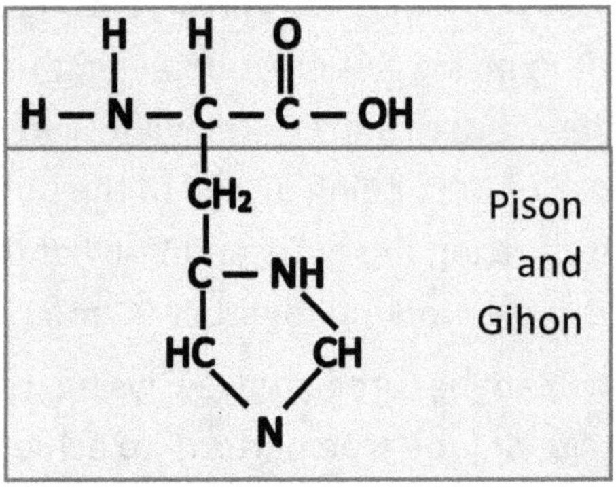


**Figure 13.15: Histidine**

with the Holy Spirit too which is consistent with the events of '*Hiddekel*'. Histidine is coded in mRNA by CA (U or C); this combines with 'G' and 'U' in tRNA, thus, all four of the 'players' are used to 'weave'. Histidine is another one of the amino acids that have to read in two ways and the nitrogenous bases reflect this.

**A summary:** For Christ centring, the Holy and the Bible (C) and Christ (A) are worked on by the Holy Spirit (the third base) to make Man and God be weaved together; Man in Christ, in the body of Christ and Christ in Man (CAU); and for Holy Spirit centring Man as a bride for Christ and the Holy Spirit in Man, (CAC). The five sided ring structure represents God in Man and for Christ centring the two 'N's (representing Man's contribution) are N3 and N4 ('*replenish and subdue*'); and for Holy Spirit centring it is N5 and N6 ('*fish of the sea*' and '*fowl of the air*').

The regenerated aspect of the born again Man is acted on by Gihon and Ethiopia and becomes glued to Christ ($CH_2$) and manifests by the three 'vehicles' ($H_3$), through two of Man's likenesses ($N_2$) and uses the four image lives of Pison' to present as joined to Christ – $C_4H_5N_2$ is the side chain for histidine, with 'a Holy Spirit switch' as appropriate.

## Threonine and Tryptophan (born-again Man gets a new image life, and all from the blood of Christ)

Threonine derives its name from a rearrangement of 'erythrose', which can be interpreted as a double metaphor for both Christ's blood (eryth) and salvation (-*ose* a sugar), respectively; this can make threonine a representative of what Christ's blood does for salvation and also salvation itself – Man representing Christ and the Holy Spirit 'sweetness' – Man a version of Christ and the Holy Spirit or Man comprising of his own 'rearrangement' by 'sugar from the blood'.

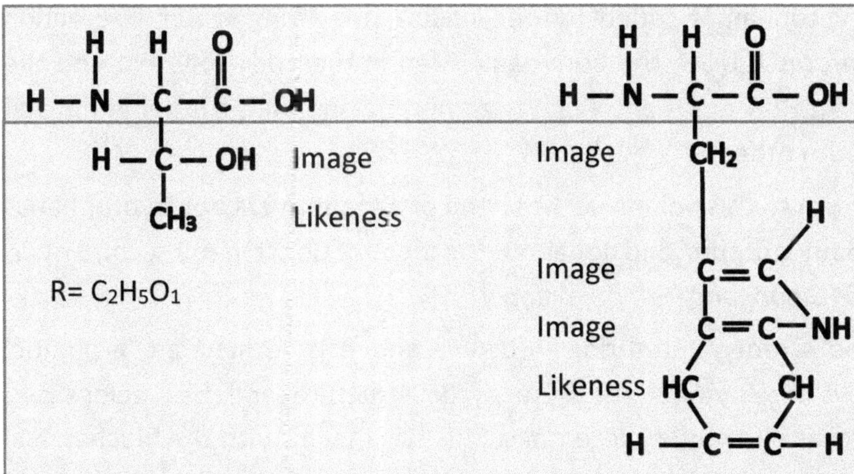


**Figure 13.11: Threonine and tryptophan**

Threonine requires all four Christ-centred interactions in order to be a version of Christ and is coded by four codons. The codons all begin with AC which is consistent with Christ like attributes coming out of Man (A) working with the Holy Spirit (C) to be like Christ. For Holy Spirit centring the emphasis switches to the 'gluing' aspect. 'The Holy Spirit selected living water interactions – one or two' apply, as appropriate.

In order to manifest as the Christ-and-Holy-Spirit-like Man, Man's whole spirit and his regenerated spirit ($C_2$) are worked on, as are the five likenesses of Gihon ('deliverers' and interactions) – $H_5$ and with the light of Christ ($O_1$); the side chain chemical formula is $C_2H_5O_1$.

The born-again Man has an interaction with living waters and the light of Christ (H-C-OH) to present as a light like Christ via 'good gold', '*bdellium*' and '*onyx*' ($CH_3$) to the outside world – a version of Christ.

Tryptophan has etymology that is consistent with an alignment with Euphrates – the born again Man (in that Man 'rubs against' Christ and the Holy Spirit) and then 'appears' as (the new Man). Euphrates demands Holy Spirit centring as this category 4 is His 'fruitful' role in Man or with Man now but the presence of Christ is needed to 'rub against'. Rather like Methionine, both Christ and Holy Spirit centring must be considered.

Tryptophan is coded by one codon (the living water interaction '*Euphrates*'), UGG, the born again Man in the body of Christ and the working Holy Spirit and Bible together making Man fruitful in himself and with others.

To get to this point Man has need of all four rivers but tryptophan is all about 'outputs' and not about the input of Christ in the second river, so no '*Gihon*' and '*Ethiopia*' apply – $H_8$.

The six-membered-ring structure above is termed an 'aromatic' ring: it is characterised by a cyclic structure and has delocalised electrons all around these rings. This ties in well with the biochemical KRM, with the 'free' electrons representing an 'active' Holy Spirit moving around the God-given-to-saved-Man image life (the 'C's). ***'Aromatic' has a meaning of 'sweet smelling'***, and this also ties in well with God describing all things to do with salvation (or God in Man) as '*sweet smelling*' (2 Corinthians 2:15[80]). The presence of these rings (generic rings) opens up a dynamic interpretation of the $C_9$:

A previously described in Chapter 12, we have 'Man's six image lives' and also a different 'God's six image lives given or made'. We can allocate the aromatic rings to either of these groups. The long-term command in Genesis is to be '*fruitful*' ('*Euphrates*'), but this can be applied both to Man himself to spiritually grow but also to evangelise with other Men. By changing the interpretation in the aromatic rings, the same splitting can be achieved – 'Man's six image lives' are all about the Holy Spirit using these to evangelise and the 'God's six image lives given' are all about Man spiritually growing with these gifts. The same can apply to the aromatic ring in phenylalanine and tyrosine. A '$CH_2$' can represent work in any 'river' that the context demands, for

[80] 2 Corinthians 2:15 (KJV): ***For we are unto God a sweet savour of Christ, in them that are saved, and in them that perish:***

tryptophan it becomes *'Hiddekel'* and *'Assyria'*; and the other five-sided ring can represent God in Man (in consistency) having two image lives of Christ and the Holy Spirit (two 'C's) and the 'N' being N7; the shared 'C's between the two rings can be representations shared by both Man and God (the *'body'* and a *'bride'*).

**Phenylalanine and tyrosine** (born-again Man shines and matures)

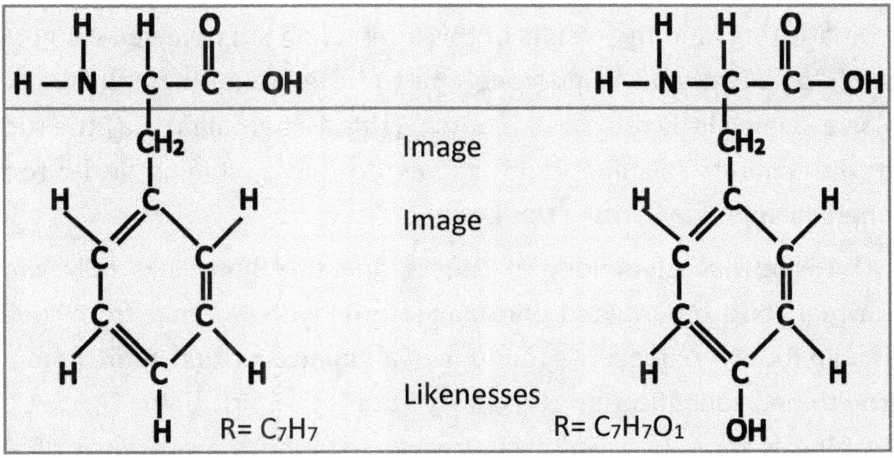


**Figure 13.14: Phenylalanine and tyrosine**

Phenylalanine has two mRNA codons UU (with U or C), and tyrosine also has two UA (with U or C). Phenylalanine can, by etymology, be interpreted as a shining rock in the church. Both Christ (John 8:12) and the Holy Spirit (the parable of the ten virgins) function to make Man a shining light, the third nitrogenous base interactions of U to A, or, C and G are consistent with these two divisions, respectively. The first two bases are U, Christ coming out of Man through his regenerated self (category 1) or Christ going into the body of Christ (category 2), and are predictable. The third base of the two codons match with specific living water interactions: (UU)U with *'Ethiopia'* and (UU)C with *'Hiddekel'*. The body of Christ changes to the bride of Christ, and the regenerated spirit to the whole spirit for Holy Spirit centring. The presence of an aromatic ring aligns with being *'born of the spirit'* and this means that the *'body'* and *'bride'* are recognised, and this makes for two more image lives being added to the standard 'four image lives of Pison' – $C_6$. To make 'Christ go in and out' of Man, all Christ-centred-water-making likenesses of the first two rivers are needed (both vehicles and interactions) – $H_7$. For Holy Spirit centring a Holy Spirit switch applies. The side chain chemical formula for phenylalanine is – $C_7H_7$.

Born again Man has an interaction with the indwelling package ($CH_2$) and further interactions with either Christ or the Holy Spirit (part of the aromatic ring) and the results of this interaction are revealed by 'good gold', *'bdellium'* and *'onyx stone'* (part of the aromatic ring) to make Man a shining little rock in the church. {This author (alan) is a little rock in the church (alanine) and strives to be a shining little rock (phenylalanine) by writing this book!}

Tyrosine has etymology of cheese and the Bible links milk with spiritual nourishment, and spiritual maturity with a change from liquid to solid food[81] – cheese. As found with arginine, spiritual nourishment, growth and sanctification is a work of the Holy Spirit (John 14:16-17). Tyrosine is thus, Holy Spirit centred but still by the presence of an aromatic ring with its meaning of completion, can be considered in a 'born of the spirit' timeline.

The chemical formula reconciliation is the same as with phenylalanine with one addition in that Man is now 'born of the spirit' and a full-of-oil-for-his-lamp Christian, hence, as this is a Holy Spirit perspective, an 'O' is added to the formula.

Tyrosine is coded by UA with U or C, and this is consistent with a first base of Christ in man being revealed in Man's regenerated spirit (the first base, A to U), and a second base of Christ being in Man (U to A). This happens in both *'Ethiopia'* and *'Hiddekel'* (one for a body and one for a bride, respectively) and tyrosine is coded by two codons.

---

[81] 1 Corinthians 3:2: ***For I fed you with milk, and not with meat, for hitherto you were not able to bear it, neither yet are you able.***

## An All-Pervasive Pattern: *'not having spot, or wrinkle, or any such thing'*

Many Christians believe that Man is inherently wicked – even the saved Man is **always** 'missing the mark'. That is not my position and would seem to contradict the Ephesians 5:27 passage in the heading above.

I believe God uses 'skin' to describe Man's status with regard to the presence of sin, and this is seen when God made *'coats of skins* (fur), *and clothed them* (sinful Adam and Eve)' of Genesis 3:21, in effect covering their sins. The Bible also describes saved Man as being able to live in the spirit (Galatians 5:15-25) and so, not to be susceptible to any lusts of the flesh. With these two things in mind, I suggest the passage above is referring to these two concepts: being sin free (*'not having spot'*) and living in the spirit (*'or wrinkle'*).[KK] I believe this is very relevant in revealing pattern in amino acid structure. In this chapter, when considering the 20 common amino acids, I concentrated on the core structure and whether it was mainly Christ centred or Holy Spirit centred; but with the broadness of the metaphoric terms used, I showed that for some amino acids it can apply to both.

You will probably be able to guess the biblical 'matching reasons' for these three divisions of 'essentiality' by now, but first let us just have another consideration of the incredible jam-packed-for-information-and-pattern Genesis 1:28, then please note that the first four gifts (N1, N2, N3 and N4) are a blessing and are just given, but the last three (N5, N6 and N7) are categorised as *'have dominion'* – this implies that Man has control over these, and, it is up to Man how far he takes them.

This brings us back to the three categories of essentiality for amino acids. I believe the 'non-essential' represent Christ centring and are a blessing (*'be fruitful, multiply, replenish and subdue'* – N1, N2, N3 and N4) and these are just given; I believe the 'conditionally essential' are Holy-Spirit centred but need spiritual growth ('fish' and 'fowl' – N5 and N6); and I believe the 'essential' is Christ centred and Holy Spirit centred ('every living thing' (N7)), but in the main sense that it is the Holy Spirit delivering, and also in the sense that He is making Man more like Christ too. Man will need a total commitment to attain them to the full, but when he does, He will be 'living in the spirit and sin free'.

Christ makes Man 'sin free' and the Holy Spirit makes him 'wrinkle free'. **(I believe the presence of 'N' in the amino acid formula is, in a way, recognising this dominion now being exercised** – if Man has attained full control, the 'N' may supplant 'H', or be in addition to 'H's if he is still growing or using His gifts on other men – the etymology will reveal where he is 'up to'.) Salvation makes you a *'bride'* for Christ automatically but you have to grow spiritually to be sufficiently bride-like (the parable of the ten virgins) and this explains why serine may be considered as both non-essential and conditionally essential.

Thus, from Christ and given to all at salvation: saved Man will; be a growing little shoot giving out light like Christ 'the herb' (asparagine) and give light (aspartic acid); a 'rock' of the church (alanine); be joined to, and in, the *'body of Christ'* (glutamic acid); and be a *'bride' of Christ* (serine) **– *'fruitful'*, *'multiply'*, *'replenish'* and *'subdue'*; Pison-and-Gihon centred; N1, N2, N3 and N4; and Christ centred (all non-essential amino acids)**.

Thus, by Man's work to grow with the working Holy Spirit: growing saved Man will be refined like silver and matured just like cheese ferments (arginine and tyrosine); he will be a vessel for giving out more and more living water (cysteine); he will give out more light and 'sweet-smelling' likenesses (glycine); he will be joined closer to image life of the Holy Spirit and Christ (glutamine); and He will be full of the Holy Spirit's fire and oil (proline). Man will grow as *'a bride'* of Christ (serine) **– 'fish'; 'fowl'; Hiddekel-and-Assyria centred; N5 and N6; and Holy-Spirit centred (all conditionally essential amino acids)**.

Thus, if Man strives and lays down his life for God to use as God sees fit, then Man will grow even more in Christ and receive Jesus Christ's gifts to the full, and then: he will be in full spiritual health (valine) and as white as snow both with what is going in and out of him (isoleucine and leucine); everything will be fruitful (tryptophan); he will be a shining rock (phenylalanine); all spiritual actions in Man will be initiated by the born-again aspects of himself (methionine); all Man's sinful promptings will be broken down by the blood of Christ (threonine and lysine); and Man, the Lord Jesus Christ and the Holy Spirit will be weaved together as one (histidine) – **'every living creature that moves'; Euphrates; N6 and N7; Man in the '*body of Christ*' and an oil-filled '*bride*' (all essential amino acids)**.

The second interpretation of the aromatic rings as Man's own image life can represent Man in the long-term aiding in the salvation of others (just as happens throughout the KRM in the Bible).

**A pervasive biochemical pattern** (deliberate biblical broad-term metaphors: deliberate broadly interpretable biochemical-etymological metaphors)

As described 'sets of six' can be considered as three 'sets of four' 'one true patterns' (two nested within one overall). But here is the incredible thing, because the biochemical etymological interpretation is also based on broadly interpretable terms, and by changing perspectives or introducing multiple perspectives, and because of chirality, then each set of six in biochemistry can also be three sets of four in the same manner (see between endnote [vii and viii]).

**The entire Calvin cycle**

RUBP (1a) and short-lived compound (1b) ∞ 3-PGA (2a) and 3 BPG (2b) ∞ G3P + glucose (3) ∞ G3P + RUBP and rubisco (4).

**The citric acid cycle**

Citrate (1A) and aconitate (1b) ∞ isocitrate + oxalosuccinate (2a) and α-ketoglutarate (2b) ∞ succinate + succinyl CoA (3) ∞ fumarate (4).

**Photosynthesis light dependent reactions** – the electron's journey (Please remember that large complexes and enzymes can represent a transaction, and by L-chirality of their amino acid components they can also represent Christ.)

Water splitting and oxygen release (1a) sunlight enters and is passed around (1b) ∞ one base pair of P680 (2a) another one of base pair of P680 (2b) ∞ pheophytin (3) ∞ plastoquinone (4).

And again:

Cytochrome $B_6$f- with 3 entities involved (1a) plastocyanin (1b) ∞ one base pair of P700 (2a) another one of base pair of P700 (2b) ∞ Ferredoxin (3) ∞ NADP to NADPH (4).

**Glycolysis** (a Holy Spirit-centred journey)

Glucose (1a) and Glucose-6-phosphate (1b) ∞ Frcutose-6-phosphate (2a) fructose-1,6-phosphate (2b) ∞ glyceraldehyde-3-phosphate and dihydroxyacetone phosphate (3) ∞ 1,3 bisphosphoglycerate (4).

And again with an overlap:

Glyceraldehyde-3-phosphate and dihydroxyacetone phosphate (1a) and 1,3 bisphosphoglycerate (1b) ∞ 3-phosphoglycerate (2a) 2-phosphoglycerate (2b) ∞ phosphoenolpyruvate (3) ∞ pyruvate (4).

**The electron transport chain**

NADH to $NAD^+$ (1a) and complex 1 gives out $H^+$ – with three entities involved in FMN (1b) ∞ $FADH_2$ to FAD (2a) and complex II pools $H^+$ (2b) ∞ complex III gives out $H^+$ (3) ∞ complex IV gives out $H^+$ (4).

**Protein synthesis**

Alanine (1a) and asparagine (1b) ∞ glutamic acid (2a) and aspartic acid (2b) ∞ all conditionally essential amino acids – matching with Hiddekel and Assyria (3) ∞ all essential amino acids – (Euphrates) (4).

**DNA base pairs; DNA to RNA transcription; mRNA to tRNA coupling**

Also all nucleotides can carry 'one true pattern' as a set of four (preceded by 'good gold', '*bdellium*' and '*onyx*':

.......nitrogenous base ∞ nitrogenous base ∞ ribose ∞ phosphate

**NAD and FAD** (as there is a shared phosphate component between the two nucleotides, the allocation of which can change in different contexts, see Figure 5.1). The nitrogenous bases will then represent the fruitful born-again Man (NAD) or fruitful born-of-water Man (FAD).

......phosphate ∞ phosphate ∞ ribose ∞ nitrogenous base

**All of these have the same underlying theme:**

***'I am the resurrection* (1), *and the life'* (2), *he that believeth in me, though he were dead* (3), *yet shall he live* (4 – *'fruitful'*)*'*;**
(John 11:25)

'One true pattern' seems to dominate biochemistry; it proclaims Christ over and over again in doing so. With this knowledge, and with the knowledge that '*world*' is often a translation of '*kosmos*' and this is a metaphor for the spirit of Man in the KRM, I ask you to consider the structure of the most famous biblical verse of all: John 3:16.

**Employed and created, not redundant and evolutionary**

As multiple codons code for the same amino acid, this is considered by some as redundancy or a 'wasteful design', and also considered by others to be a 'chunky' design; *but now we know that just ain't so – it's designed to get the correct number of participants and interactions in order to testify to Christ, multiple times, over and over again.* Others point to the same compounds being used across plants and animals, or in different processes within the same organism, and they say these point to shared evolutionary predecessors and processes; *but that ain't so either – it's designed to have the same consistent metaphoric meaning throughout the biochemistry of all life in order to testify to Christ, multiple times, over and over again.*

Christians have long been '*defenders of the faith*', but I now believe the whole argument will change and Christians can now have an 'offensive game' too: *Explain this shared pattern then! If you can't, then there must be a God and He is the God of the Bible; the God of Abraham, Isaac and Jacob; but will you acknowledge and serve him?*

**Ochre, amber and opal**

Then we come to the three stop codons and unusually, these have actual nicknames, and the names are based on colours, UAA (ochre), UAG (amber) and UGA (opal); and we have already seen the significance of colours in the biochemical KRM. Stop codons do not pair with an anticodon but cause the release of the synthesised peptide 'out into the environment'. In the rivers of Eden passage, there are also three entities that are the source of saved Man's gifts that are released into the environment: 'good gold', *'bdellium'* and *'onyx stone'*. These in effect, are the interface between Man and the outside world.

***Ochre has etymology associated with an original meaning of 'pale yellow (gold)'***, but the meaning has changed over time to include a representation of a 'natural earth pigment associated with iron oxide and ferric oxide'. This expands its colour meaning to include orange and red (the whole can be interpreted to represent a version of Christ as 'gold' and Christ's blood sacrifice 'mixed' together).

***Amber,*** although used as a colour for naming the codon, is a consistent metaphor for 'what comes out of the Holy Spirit'.

***Opal has etymology of a 'precious stone' and specifically 'a change in colour'.*** This aligns well with *'onyx stone'* as a representation of the born again, changed in 'colour' Man (*'scarlet'* red to pale pink as found with 'onyx stone', or even white)

Thus, the three stop codons align perfectly with 'the three resources' of 'good gold', *'bdellium'* and *'onyx stone'*. As well as this, the order of nitrogenous bases describes born-again Man from these three perspectives in 'one true pattern' format:

UAA (ochre): the regenerated Man is in the body of Christ and gives out Christ-like attributes.

UAG (amber): the regenerated Man is in the body of Christ and gives out gifts of the Holy Spirit.

UGA (opal): the regenerated Man is glued to Christ by the Holy Spirit and together give out the light of Christ.

Elegant summaries of the salvation process are revealed and image-to-likeness order is maintained and this now means that all codons and their nitrogenous bases are consistent with the Genesis text.

**Meaningful mutations: design after all?**

I will show later in this book how the diversification of species and species adaptation testify very specifically to spiritual gifts given to saved men through salvation by Christ. This process is driven by supposedly random mutations (I prefer to call it refinement and inbuilt genetic potential being realised). Once we know that nitrogenous bases have analogue meaning then mutations may also have a metaphoric meaning and by this second meaning they can reveal design and confirm my position. As one example, the commonest mutation is a single G-T mutation (Holy Spirit (G)-original spirit of Man or Jesus as a man (T)), G is replaced with T in the mutation.

If I read these pre- and after mutation nucleotides together then (G to T) can be interpreted as: the Holy Spirit works on Man, goes into Man or presents Jesus as a man. In a deleterious way it can be interpreted to mean the Holy Spirit is replaced by Man. This 'mutation' can be bought on by the presence of 'reactive oxygen' (radicals derived from oxygen that have one or more free electrons); we now know that this can represent the Holy Spirit working with 'light' and this is consistent with the Holy Spirit washing Man with light (G-T). Acidic conditions (increased $H^+$ ions) can also increase mutations and I suggest that all mutation processes may reveal design.

Romans 8:22 tells us that the '*whole creation groaneth*', not just some of it but all of it, I suggest that this includes light speed, zero-point energy, atomic energies and DNA. DNA mutations may be the 'creation groaning' or they may be a design that represents changes in Man's spirit after salvation; hence they may be beneficial or harmful.

## Cancer of the *'old self'*

The biochemical KRM is all about biochemistry mimicking the salvation process; Man's biochemistry moving forward metaphorises Man's spirit before, at and after salvation. Cancer can be considered as fundamentally a disease of DNA and has significant parallels to being a representation of Man's sinful self (called his *'old self'* or *'old man'* in the Bible), and this *'old man'* fights salvation changes.

***Cancer has etymology associated with a crab because it was noted that tumours invaded tissue 'like crab's legs' and 'grabbed' their surrounding tissue***. And this is consistent with a representation of the sinful nature of Man invading and breaking down his salvation gifts. Cancer tissue metabolises and breaks down all amino acids and so, in parallel Man's sinful nature attacks, grabs and breaks down all the gifts that we now know amino acids metaphorise.

In particular many tumours break down methionine and as this represents how all Man's post-salvation gifts are initiated then Man's *'old self'* will readily stop this occurring. Likewise, glutamine (the Holy Spirit's work joining Man to Christ) is another primary driver for tumour growth and this also parallels the *'old self'* breaking down this work-of-the-Holy-Spirit gift also.

Tumours seem to break down more Holy Spirit centred- or Man's-own-personal-growth-centred gifts (all represented by amino acids) as opposed to Christ-centred ones; and this ties back to the use of *'dominion'*, it is up to Man how far he takes gifts that are under his own dominion.

Thus, the amino acids considered to be the most consumed by tumours are glutamine, methionine, leucine, isoleucine, serine, glycine, asparagine and valine (asparagine is the only Christ-only centred amino acid featured but it specifically represents the individual persons growth).

**A summary of the rules of the metaphor so far**

- A name's etymology will introduce you to an existing biblical metaphor (like malate and apple); or a specific role of an entity like (adenine and Christ both 'giving'); or sometimes a specific trait (oxaloacetate and Christ-like attributes).
- The chemical formula will confirm the metaphor and reveal what the entity is working on or with; sometimes the chemical formula can be exactly predictable from Genesis (nucleotides).
- Acid-to-base conjugation (and deprotonation in general) will parallel forward progress in the salvation process (like pyruvate). Electrical-charge change will reveal what the Holy Spirit is working on and reveal forward progress in the salvation process (like phosphate groups).
- Sequential compounds in processes will reveal forward progress in the salvation process (like Krebs cycle).
- Large compounds can be broken down and moving between these components will reveal forward progress in the salvation process (amino acids and DNA).
- Compounds in biochemical and physiological processes will reveal a chronology that matches the Genesis text (like glycolysis and photosynthesis with Genesis Day 4).
- I am about to show how the environment, cross-organelle- and cross-organ-biochemical processes, and physiological systems all parallel salvation (photorespiration and haemoglobin).

For those interested in research my 'miss rate' was zero, every single compound I looked at had Genesis alignment – over 50 so far and with more to come. Not only that, but the chronology of every reaction, even in the most complex and complicated systems, were an exact match for the chronology of the biblical metaphor (or the biblical text itself) and all of it is within a 'one true pattern' framework.

This is why I claim proof because what are the chances?

# Chapter 14

# The Body, a Bride, Porphyrins and the Purple-Robed Lord Jesus Christ (II)

*Genesis-day-aligned biochemistry (VI)*

Genesis 1 Day 6, in the KRM, defines the *'body'* and a *'bride' of Christ'*:

> **Genesis 1: 24-25 (KJV):** ***And God said, Let the earth*** *(eres)* ***bring forth the living creature after his kind, cattle, and creeping thing, and beast of the earth*** *(eres)* ***after his kind: and it was so.***
>
> ***And God made the beast of the earth*** *(eres)* ***after his kind, and cattle after their kind, and every thing that creepeth upon the earth*** *(adamah)* ***after his kind: and God saw that it was good.***

As a recap, please note the differences between the two verses: the *'living creature'* disappears in the second verse; there are changes in which categories are classed as a *'kind'*; the *'creeping thing'* becomes *'a thing that creepeth upon the earth'*; the *'beast of the earth'* becomes *'**the** beast of the earth'* and also changes from being the last-named animal life to the first-named; and there is a different root word change to *'adamah'* from *'eres'* for *'earth'*.

There are very precise reasons for all of these changes, but I am going to focus on the *'bride'* and *'body of Christ'* and show how they fit into biochemistry. The *'body of Christ'* is this mysterious concept of all saved men forming a 'body' – both their image life and likenesses are a part of this body, and the body is centred on Christ's image life given to Man; the *'body of Christ'* as a whole is classed as image life – so likenesses of Man become an image (likenesses are skillsets, traits, behaviours and an identity). Jesus Christ is the head of the *'body of Christ'*. *'Beast of the earth'* is how God metaphorises all of Man's likenesses given at birth; and these are made more Christ-like and also added to at salvation (and after that). The Genesis text has an image-to-likeness structure, and so by changing *'beast of the earth'* from the last-named animal in Verse 24 to the first-named in Verse 25, and also by adding a definite article ('the'), this satisfies the KRM (and 'one true pattern') in representing image-to-likeness structural changes at salvation with the formation of the *'body of Christ'*.

The use of *'cattle'* is a way of describing something living 'under the shelter of someone else' (like domesticated animals do). It may refer to the indwelling of Jesus Christ and the Holy Spirit placed in Man, or may mean Man's original spirit and his regenerated spirit being in the *'body of Christ'*, or a *'bride'* for Christ, and it may even refer to Christ and the Holy Spirit in God the Father in their triune status. A *'bride'* for Christ is also classed as image life: the *'bride'* is a personal relationship perspective but now centred on the whole spirit – how close to Jesus you become and how much of the Holy Spirit you can carry (both qualitative), whereas the *'body'* is either in or out (quantitative).

The Bible likens building up a relationship with the Holy Spirit to ten virgins having 'sufficient oil level in their lamps' (the parable of the ten virgins – Matthew 25:1-13). Thus, the biochemical KRM may use the 'storage' of biochemical compounds that represent Christ and the Holy Spirit (like oxygen, myoglobin or glucose). These, together with chirality, are both used to align with Day 6 (the *'bride'* of Christ may even be considered a subset of the body, as all the members of the 'bride' (or all brides?) are in the body but not necessarily vice versa – as the parable above suggests you need to have 'sufficient oil').

## The Theory: A Spatial 'One True Pattern'

When considering light-dependent reactions in photosynthesis we established that the intra-organelle space (in thylakoids in chloroplasts) was called the 'lumen', and this was where God's light was now in Man (a regenerated Man); this is adjacent to the intermembrane space where the interaction to place God's light in man took place (Christ going into Man); and adjacent to this was the 'stroma' or outside world where the Holy Spirit with or without the Bible interacts with the inter-membrane space. Thereby a spatial 'one true pattern' is set up – the place where one true pattern exists:

Regenerated Man ∞ Christ in Man ∞ the Bible ∞ the Holy Spirit

Luman ∞ intermembrane space ∞ stroma ∞ stroma

The same happens in the electron transport chain:

Matrix ∞ inner membrane ∞ inter membrane space ∞ inter membrane space

(I also repeat-speculate that this is consistent with 'phospholipid membranes' representing God the Father (lipid) as a 'light bringer' (phosphate) and everything enclosed within them is His purpose.) The same spatial pattern happens in photorespiration but this time across different organelles in plant cells.

### Photorespiration

We discussed photosynthesis non-light-dependent reactions, and briefly mentioned photorespiration, in Chapter 10. Ribulose 1,5-bisphosphate and rubisco together fix carbon dioxide from the atmosphere into organic compounds. However, if there is a lot of oxygen in the vicinity, they react with the oxygen instead – this is called 'photorespiration'. Most biologists consider this to be a detrimental reaction when considering the efficiency of photosynthesis, as it uses a lot of energy. (I do not believe this, and I think there will be other 'reasons' for photorespiration!) Photorespiration is a very complicated series of reactions and carries many parallels with the washing clean of Man's spirit by the light of Christ and by the Holy Spirit ($O_2$), and it also parallels the creation of the '*body of Christ*' and a '*bride*' for Christ. Rubisco (Christ by chirality) and ribulose 1,5-bisphosphate (the Holy Spirit by chirality and etymology) can work with either carbon dioxide (Man's spirit) or oxygen (any of the Trinity-God's light).

***Peroxisomes are organelles in plant cells, and the name has an etymology stemming from 'hydrogen peroxide' and 'body' (some);*** 'hydrogen' means 'water making': and 'peroxide' has a meaning of

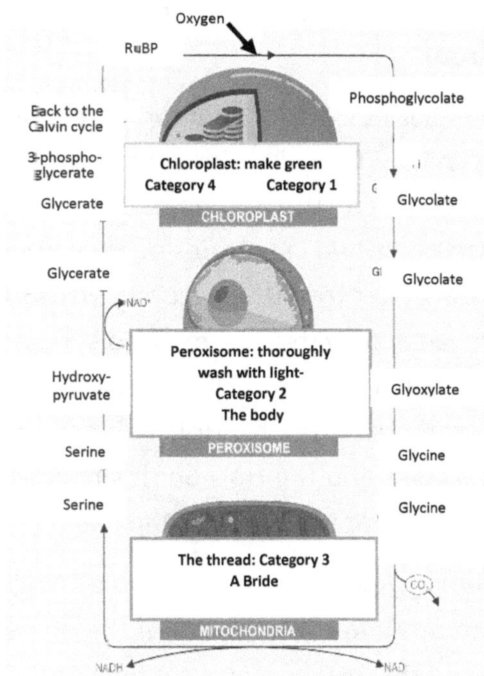


**Figure 14.1: Photorespiration; Ant animation/Shutterstock**

'thoroughly oxygen'. The whole is consistent with representing Man's spirit and waters being made and filled with Christ's or the Holy Spirit's light ('thoroughly oxygen') within the '*body of Christ*' ('some').

Another spatial 'one true pattern' is apparent (Figure 14.1) but with a 'looping' of the '*bride*' back to the '*body*' (the '*bride*' in the '*body*').

## The Players, the Proof and the Pattern

In the biochemistry of photorespiration there is now a move to $C_2$ compounds and this is consistent with representing a focus on some specific aspects of Man's spirit and not the whole spirit or whole process, and in particular: his original spirit; his regenerated spirit; and how these form Man's *'beast of the earth'*, the *'body'* and *'bride'* of Christ – a 'born of the spirit' view. I am again, only presenting a simplified version of photorespiration here but please note in photorespiration, compounds are taken from one organelle to another and their element component's metaphoric representation may change with this movement (often a change from Christ centred to Holy Spirit centred).

Because a smaller number of component elements are considered this allows for more flexibility in allocation of their metaphoric meaning. Also, several of the compounds can be interpreted to be Christ centred (the first two categories of 'one true pattern') and both Holy Spirit and Christ centred (the second two). The Christ-centred $C_2$ for a conjugate acid may represent Man's regenerated spirit and Christ's image life, and for the base it may represent *'the body of Christ'* and Man's *'beast of the earth'*; for Holy Spirit centred it may be Man's whole or original spirit and the Holy Spirit's image life in the acid, and a *'bride'* and the *'beast of the earth'* in the base.

There is generally more than one plausible 'allocation' that will lead to the desired count, and this is for all of the element categories ('C', 'H', 'O' and 'P') and this makes it less confirmatory.

We have already covered the etymology of 'gly' (glycol) as being associated with salvation itself or of 'God in Man'; and 'phospho' as being a light bringer. We also have the consistent use of numbers as representing one of: image lives, likenesses, living water or lights; and many of these 'sugar and light' etymologies are broad and flexible terms that can be applied to different entities in different timelines.

**Glyceraldehyde 3-phosphate (G3P)** $C_3H_7O_2PO_4$ is D chiral.

**2-phosphoglycolic acid** ($C_2H_5O_2PO_4$) – 2-phosphoglycolate ($C_2H_2O_2PO_4^{3-}$) is achiral.

**Glycolic acid** ($C_2H_4O_3$) – **glycolate** ($C_2H_3O_3^-$) is achiral.

**Glyoxylic acid** ($C_2H_2O_3$) – **glyoxylate** ($C_2HO_3^-$) is achiral.

**Glycine and glycerine** both have an etymology of 'sweetness'.

**Serine** has an etymology of 'silk-like' and is L-chiral. The Bible links silk with splendour and God's provision and also with women (a *'bride'*).[82] Please note, in Ephesians 5:23-30, the comparisons of a *'body'* and a *'bride'*; how a *'bride'* is washed by the Word of God and is considered a part of the *'body'*. [83,84,85]

**Hydroxypyruvate** has an etymology associated with pyruvate (the Bible as 'fire' and the 'grape'), but with more waters and oxygen: it can be interpreted as the Bible 'washing man with the Word of God'.

***Ketoglutarate and glutamate*** have this shared etymology of 'gluing' (to Christ individually or as a part of the 'body of Christ'), as discussed in Chapter 12 for the citric acid cycle (for ketoglutarate and not glutamate). Note that glutamate is L-chiral and ketoglutarate is not. The change of ketoglutarate to glutamate can be interpreted to mean Man's becoming a part of the *'body of Christ'* and a *'bride'* for Christ.

**The versatility of Genesis 1:24 and 25** (*'cattle'* is plural by *'their'* kind)

Man (living creature – Category 1) ∞ Christ indwelling (one of the cattle – Category 2) ∞ the Holy Spirit (one of the cattle – Category 3) ∞ the Holy Spirit and Bible (creeping thing – Category 4)

---

[82] Proverbs 31:22 (KJV): ***She maketh herself coverings of tapestry; her clothing is silk and purple.***

[83] Ephesians 5:23 (KJV): ***For the husband is the head of the wife, even as Christ is the head of the church: and he is the saviour of the body.***

[84] Ephesians 5:24 (KJV): ***So ought men to love their wives as their own bodies. He that loveth his wife loveth himself.***

[85] Ephesians 5:26 (KJV): ***That he might sanctify and cleanse it with the washing of water by the word,***

The ‘cattle’ and ‘creeping thing’ have ambiguity and so can represent the Holy Spirit plus or minus the Bible. (Please also always bear in mind the circular nature of the ‘one true pattern’.)

This is God’s perspective, this is what God (*‘earth’*) brings forth, but a Man’s perspective is also present as he too can be represented by ‘earth’: Man (earth) brings forth *‘the living creature’* now representing a regenerated Man (1); as *‘cattle’*, a presence in the *‘body of Christ’* (2); as *‘cattle’* a presence as a *‘bride’* for Christ’ (3), and *‘beast of the earth’* will become Man’s fruitful likenesses (4).

In Chapter 2 we discovered how this was an image count for thymine and Verse 25 was an image count for Uracil. Verse 25 shows a similar pattern but this is now God centred only (‘God made’). By the use of *‘adamah’* and by the use of a plural for *‘cattle’* then that plural may be two or three indwelling entities. This means that *‘cattle’* can contain Christ, the Holy Spirit and the Holy Spirit and the Bible, and this means *creeping thing that creepeth upon the earth (adamah)* can be looked on as a long term instruction to work on a changeable part of just Man’s spirit and not the whole world (*‘eres’*). Thus, the KRM for this day emphasising *‘born of the spirit’* centring and not evangelising.

Photorespiration seems to adhere mainly to the Man-centred version of Genesis 1:25, and in so doing it shows how photosynthesis covers the entire salvation of Man as described in the KRM – Genesis Days 4 to 6.

**Photorespiration in the KRM** (Please see figure 14.1)

**Chloroplast: Green Herb: category 1: *the beast of the earth*** – rubisco (Christ by chirality) and ribulose 1,5-bisphosphate (the Holy Spirit by chirality and etymology) metabolise oxygen (work with more of Christ’s light on thriving-Christian-Man’s spirit) rather than directly with carbon dioxide (Man’s spirit).

During this representative oxygen metabolism, in the biological KRM, the Holy Spirit still uses Christ’s light to work on the regenerated

Man (3-phosphoglycerate). The light-bringer-regenerated Man gives out light in his light bringer status (phosphoglycolate).

**Peroxisome: category 2: *'Cattle'* (Man in the body of Christ)** – glycolate is toxic and washed clean by oxygen to form hydrogen peroxide and glyoxylate. Man's whole spirit is washed clean by the light of Christ, and this aspect is represented by glyoxylate. Some glyoxylate becomes a part of (glued to) *'the body of Christ'* (there is a concomitant glutamate-ketoglutarate reaction here – which consistently signifies being one with the *'body'* as found in the citric acid cycle). The glutamate may also be a reference to the 'gluing' of the 'cooperative'.

**Peroxisome: category 3*: 'cattle'* (Man as a bride for Christ)** – Man's original spirit gets washed clean by the Holy Spirit – glyoxylate – it is achiral (glyoxylate has $H_1$ in its formula and this is consistent with a representation of 'light' in general, glyoxylic acid has $H_2$ and this is consistent with a representation of '*Hiddekel*' and '*Assyria*'). Man gets living-water-making capabilities from the Holy Spirit – glycine.

**Mitochondria: category 3: *'cattle'* (Man as a bride of Christ)** – Man gains more light from the Holy Spirit (serine and glycine). Man becomes a *'bride'* of Christ (serine). A concomitant $NAD^+$ becomes NADH here and confirms an addition of light and living waters to Man's spirit – this light is now from the Holy Spirit.

**Peroxisome: category 3: *'cattle'* (as a bride for Christ)** – The *'bride'* is a part of the *'body of Christ'*, and both are washed with the *'water of the word'* (hydroxypyruvate).

**Chloroplast: category 4; *'thing that creepeth upon the earth'*** – Man's 'sweet likenesses' (glycerate – the same consistent metaphor as previously used in glycolysis, but without a light bringer or the Holy Spirit) continue to work with the Holy Spirit and re-join the Calvin cycle to *'be fruitful and multiply'* in the world. There is a concomitant NADH becoming $NAD^+$ here and this confirms Man giving something in this phase, and a concomitant ATP becoming ADP confirms God's light is added to Man by the Holy Spirit – *'ereb to boqer'*.

**A pigmented Gospel and haemoglobin (*'the blood sphere'*)**

And finally, back to haemoglobin: not only does the KRM cross plant types (CAM, C4 and C3); parallel whole processes (cell respiration and photosynthesis); join mechanisms (electron transfer chains); but also, the Kosmos-Ruah Metaphor is 'told' by singular compounds (or parts of compounds). Heme-containing proteins do just this.

In the biblical KRM, our atmosphere (called a *'firmament'* or when populated a *'heaven'* in the Bible) represents the indwelling of God in Man (image life that includes the '*body*' and '*bride*'). Thus, even within the KRM, our atmosphere may be both literal, representing the whole world, and metaphorically representing an indwelling. At the same time, throughout Genesis 1, the body of Man metaphors his spirit (*'let us make Man'* meaning organic Man but also remake Man's spirit in the KRM). With these representations in mind:

Haemoglobin transports oxygen from the atmosphere and then lungs to Man's tissues – and in parallel, light (O) goes from the indwelling to Man's spirit. Man's body produces carbon dioxide and this is excreted into the atmosphere. Man's spirit can be considered as an image life and two lights ($CO_2$) and it is 'excreted' (presented) to the whole world as a 'light of the world' (atmosphere then has a literal meaning of the whole world). Alternatively, Man's born-again spirit (represented as an image and two lights again ($CO_2$), is placed in the body of Christ or is a bride for Christ (atmosphere now has metaphoric meaning of an indwelling). There can be the usual image and lights changes for Christ centring or Holy Spirit centring for the $CO_2$.

If we consider haemoglobin as Christ's blood and we know that the interface between Man's indwelling and Man's spirit is all by Christ's blood then Christ can be inserted into the gaseous exchange in the above paragraph: Man's organic body (Man's whole spirit) – haemoglobin (the interface with Christ) – haemoglobin (the interface of the Holy Spirit) – the atmosphere (Man's indwelling).

Haemoglobin has a quaternary structure; each 'quarter' or subunit is composed of peptides, each with an iron-based heme. There are two $\alpha$-globin peptide chain subunits and two $\beta$-globin peptide chain subunits, this means a 'one true pattern' can readily be established:

Alpha globin 1 ∞ alpha globin 2 ∞ beta globin 1∞ beta globin 2

(At any one time), I will allocate 'alphas' as Christ centred and 'betas' as Holy Spirit centred. The transported oxygen atoms are bound to iron atoms in each of the hemes. Man breathes in and oxygen is taken into the lungs and transported via haemoglobin to the tissues. Christ's light or the Holy Spirit's light (oxygen) from Man's indwelling (Heaven) moves into the tissues of Man (Man's spirit) in all four one-true-pattern categories. By this light Man's spirit has a changed 'image and two lights' ($CO_2$) and this is given out as light to the whole world (in categories 1, 3 and 4) or back into making Man's spirit in category 2 – the atmosphere flipping in meaning accordingly. The 'image and two lights' ($CO_2$) change from regenerated Man and the greater and lesser lights, to the whole spirit of Man and the Holy Spirit and Bible as lights, again accordingly. Category 3 can also flip to mean the Holy Spirit's indwelling presence grows as with Category 2 for Christ.

The carbon dioxide bonds to an entirely different place than the oxygen and actually binds to the amine group ($NH_2$) of a valine amino acid in the peptide chains. From Chapter 13, we can say that this is consistent with the transported $CO_2$ representing an image and two lights becoming a part of the 'indwelling' (our atmosphere) – making Man more Christ- and Holy Spirit-like; having more 'health' (valine).

I think Deuteronomy 12:23 '...***blood is the life***...' will prove to be very significant in the biochemical KRM, as there are so many layers to haemoglobin function that parallel salvation: blood buffering; nitric oxide (NO) transport; and concentration gradients. Here is an example of just one: the $CO_2$ binding described above happens when there is a lot of oxygen (Christ's light or the Holy Spirit's light) present but when there is not, then arginine and lysine in the haemoglobin peptide chains bind $H^+$ (light) that stabilise and enhances the $CO_2$ binding. This can represent Man's sinful spirit being broken down (lysine) and refined (arginine) to become more of the *'body'* and a *'bride'* when Man is short of 'light' so that he can acquire more light. I think blood will need volumes of books written about its parallels with salvation!

Each heme has four pyrrole rings (see figure 3.1). These rings have five vertices, four of carbon ($C_4$) and one of nitrogen (N) and a central iron core (the iron is symbolising that all of the metaphoric meaning is the result of Christ's blood sacrifice). I suggest the 'N' representation may change depending on context but is often N7 – Man's spiritual growth and evangelisation. The carbon count of these rings will represent the image-life 'players' and in the first two Christ-centred categories of the 'one true pattern' they are: the four image lives of 'Pison' – $C_4$. For the second two Holy Spirit-centred categories, the regenerated spirit drops out and the whole or original spirit is added. As well as this, each pyrrole ring, as a whole, will also represent any four from: God the Father, Jesus Christ, the Holy Spirit, the Bible or regenerated Man, again depending on context and timeline.

Another heme, this time in myoglobin, stores oxygen in muscle tissue for use in muscle activity or under duress, just as, in parallel, *'the body' and 'bride'* of Christ increase 'the oil in their lamps' or Christ's light in themselves and store it. Myoglobin also exists in large quantities in heart tissue, just as Man 'stores' his 'light' in his heart (the 'heart' is a term used biblically for the spirit and soul). Interestingly, whales also have huge amounts of myoglobin, and this seems fitting as I will show that they also metaphorise Christ's indwelling in the KRM. An elegant metaphor is thus produced and I find the concept of blood representing Christ's interface with, and in, Man appealing. **The red-coloured heme reveres Christ's precious blood sacrifice.**

**The colour purple** (porphyrin)

Plants absorb mainly red and blue light, and red and blue make for purple. Plants also have a green-coloured chlorophyll porphyrin, just like Christ is likened to a *'green herb'* in Genesis 1:30 and probably in Luke 23:31. 'Plastocyanin' presents the 'blue bridge of Christ' to Man; put that with the red heme and mix it up, and get the purple-robed Lord Jesus Christ – our saviour.

**God is light**

I suspect the metaphor is not just in biochemistry and physiology either; *'God is light'* declares 1 John 1:5, and God Himself contains Jesus and the Holy Spirit within Him as part of their triune nature. Physical light also has a wave-particle duality within it: I wonder if the wave is representative of the Holy Spirit and the particle, of Christ (others have suggested this too). The same is true for the apparent wave-particle-like duality for all atomic components. Now that we know the electron metaphorises the Holy Spirit, it could make the more particle-like components (protons and neutrons) represent Christ.

Perhaps zero-point energy is representing the Holy Spirit too and maybe 'quanta' are the units of His activity or maybe they represent God the Father 'purposing'. The omnipresence of the Holy Spirit may even be represented by the 'probabilistic nature of particle location' (they are 'sort of' anywhere you look for them, as is He!).

**A summary: anyone can make up parallels**

I am accused of making up parallels! *'Anyone can see things if they look hard enough.' 'Anyone can say, 'let so-and-so be so-and-so' and make up almost any story they like.'* Yeah, really? Show me. Find a book yourself and use 56 consecutive sentences (Genesis 1 and 2 have 56 verses) that have a hidden repeating metaphor (at least 15 times) that describes the rest of the book; then, using etymology and logic, show me that same story throughout the biochemistry and physiology of all life. Good luck with that – as I have tried to see the salvation story in other books, tried to find other non-biblical stories in Genesis, and tried to see similarities in biochemistry with other narratives too, and I cannot even get started – the etymology alone is too hard to get past.

I find the nucleotide chemical formula calculations easy to follow and almost irrefutable as proof of a shared design on their own; the amino acid and other compound counts sometimes have more options at coming to the 'correct' figure, but the fact that they can all be done is still strong evidence of design for me. Couple that with the perfect chronologies (these are the most compelling for me) and the fact that this (the Bible) is written before any biochemistry was known.

I think, even knowing biochemistry, it would be very difficult to actually write a coherent book with biochemical reactions hidden in it, especially if you had to tie it to etymological meanings, and all in 56 sentences. **Of course, if you did write such a book (by design) it would, in a way, confirm my claims – that the Bible and biochemistry share the same design (by design) and it must be by the same designer.**

When I used to read Genesis 2 and the rivers of Eden passage, I always used to wonder why they were included. Strangely named rivers, destinations and resources, what have they got to do with anything? I used to skip it out a bit! But it turns out that they are also a 'big deal' and are actually pivotal in defining the biochemistry of all life on Earth; they reveal an even 'bigger deal' in describing the salvation of men: and these, in turn, all testify to the '**very biggest of all big deals' – the King of Kings – the Lord Jesus Christ Himself.**

***'... everything was created by him and for him' (Colossians 1:16)***

**Confirmed ✓**

# Part 2

## The Two Worlds

**Genesis 1 and 2 clearly describe God making two worlds on our Earth.**

**The first world represents Adam's spirit and how all men were meant to be, and it reflects Jesus Christ as God in Heaven.**

**The second world, the world we live in now, represents the spirit of all saved men and reflects Jesus Christ who became a man and came down to Earth.**

**The way both of these worlds are illuminated; the water cycles; the interfaces between land, sea and air; the size of trees; animal adaptation and diversification; and the size, distribution and type of animal life will all have a 'biblical reason'.**

**The Genesis six-day account is a story of God remaking the first world, and also, by metaphor, a story of God remaking Man's fallen spirit – born-again world and born-again Man – the Trinity God's individual roles are the same for each.**

**With this knowledge, an entirely different explanation for the entire fossil record becomes readily apparent.**

# Chapter 15

# Gap Theory

# (But Not as We Know it)

## *Biblical evidence of a gap*

Gap Theory is defined as follows: God made two worlds, and the Genesis six-day account is the story of God remaking a previous world; Genesis 1:1 is referencing God making that first world; Genesis 1:2 is a description of that world after judgement; and Genesis 1:3 onwards is all about the remake, the world we now live in – there is a time gap of indeterminate length between Genesis 1:1 and Genesis 1:2.[86]

I have never liked the name 'Gap Theory', as it places too much emphasis on the gap, which seems to be of little metaphoric meaning (although I suspect the gap duration has a factor-of-three relationship with the duration of life in either world – this, again, would then be a reverence to Jesus Christ's three days in the grave). I have come up with some new names which I will share later.

The biblical evidence for a gap is overwhelming:

---

[86] Genesis 1:1-3 (KJV):

***[1] In the beginning God created the heaven and the earth.***

***[2] And the earth was without form, and void; and darkness was upon the face of the deep. And the Spirit of God moved upon the face of the waters.***

***[3] And God said, Let there be light: and there was light.***

***[4[ And God divided the light from the darkness.***

***'It was so'* obvious:** *'it was so'* comes after the 'actioning' (*'God made'*) statement in Day 2 (after God made the firmament), in contrast to Days 3, 4 and 6, where it is placed after the 'command' statement (*'God said'*). There are specific reasons for this difference discussed in the KRM book, but, I suggest, in Days 3, 4 and 6 (as Genesis is generally chronological throughout) the *'it was so'* statement cannot be a description of God's command being actioned; it is not actioned until the *'God made'* statement. Some argue that *'it was so'* just confirms the authority of what God said, meaning 'it was going to be done'. **I suggest it is a plain-text, literal and obvious confirmation of Gap Theory, referring to life or the Cosmos in the previous creation**. In other words, ***'it used to be that way'* (in the first creation).** Simultaneously, in the KRM, it describes a status of, or actions on, Man's fallen spirit – 'it used to be that way' (before salvation) – a classic hidden in plain sight if ever there was one.

To help to confirm this: *'it was so'* is never used when Man is the subject matter or when the word 'created' (*'bara'*) is used. Notably, on Day 5, *'it was so'* is missing. In that day and in Day 6, *'bara'* is used for Man, whales, birds and some other sea creatures – all of which, I suggest, are not in the original creation. This is consistent with *'bara'* meaning a new creation from God and also explains the omission of *'it was so'* from any statement referencing Man or created creatures – **if they are a new creation or Man, they cannot be 'it was so', as they are not there in the first creation**. This is most notable in Genesis 1:29-30 where Man is proclaimed to be a vegetarian (for that time being) and does not get *'it was so'* attached, but plants do in the next verse.[87] I know of no other reason for the selective use of *'it was so'*.

---

[87] Genesis 1:29-30 (KJV): [29]... ***and every tree, in the which is the fruit of a tree yielding seed; to you it shall be for meat.*** [30] ***And to every beast of the earth, and to every fowl of the air, and to every thing that creepeth upon the earth, wherein there is life, I have given every green herb for meat: and it was so.***

**Genesis 2:4:** This is my favourite verse in the Bible, in that there is so much information crammed into it, and especially the way it links with Genesis 1:1 and Genesis 1:27.

> **Genesis 2:4 (KJV):** ***These are the generations of the heavens and of the earth when they were created, in the day that the LORD God made the earth and the heavens,***

You really do have to stretch this text to see anything other than that there is more than one '*generation*' of an '*earth and heavens*'. '*In the day*' God made the Earth, there is more than one generation of it – simple, there was a previous world. Please note the details: 'heavens' is plural and 'earth' is singular; the first world is created and the second made: 'heavens' and 'earth' matches with the order of Genesis 1:1, and 'earth' then 'heavens' matches Days 3 and 4. There are more very precise reasons for these details to come.

**Genesis 1 is sequential:** Genesis 1:1 describes a created and completed world, and everything else is after that. This can only make sense with two worlds.

**'Tohuw'**: Isaiah 45:18[88] describes how God does not create anything 'tohuw' yet the created world (past tense) in Genesis 1:2 is described as 'tohuw' ('*without form*'). The same word – something must have happened to it to make it 'tohuw'.

**Jeremiah 4:23-25**[89] describes how a previously fruitful place was now a desolate and destroyed world with no humans, stars, Sun or Moon.

---

[88] Isaiah 45:18 (KJV): ***For thus saith the LORD that created the heavens; God himself that formed the earth and made it; he hath established it, he created it not in vain, he formed it to be inhabited: I am the LORD; and there is none else.***

[89] Jeremiah 4:23-25 (KJV): ***I beheld the earth, and, lo, it was without form, and void; and the heavens, and they had no light... I beheld, and, lo, there was no man, and all the birds of the heavens were fled.***

**Willingly ignorant:** 2 Peter 3:5-7[90] describes men being wilfully ignorant (please do not be in this people group) about a flood, and also describes two worlds (kosmos) *made* 'by the same word' (Jesus Christ).

**Replenish**: Noah is told to replenish (*'mala'*) the destroyed world after the flood (Genesis 9:1[91]). God uses exactly the same word in Genesis 1:28[92], and this is consistent with a world being remade – you cannot replenish something unless it was originally stocked.

**Renewest:** Psalm 104:30[93] refers to God 'renewing the face of the earth'; this may be alluding to seasonal changes or Noah's flood but may also be an allusion to a second world.

**A quickly removed flood:** Psalm 104:6-7[94] describes a flood covering the whole world that was removed quickly (like the waters in Genesis 1). This cannot be Noah's flood, as that dissipates slowly.

**Job 38** describes water breaking forth quickly like waters breaking in pregnancy (unlike Noah's flood, which was over 40 days), and it also describes seas being frozen like a stone. I will show how both of these descriptions 'fit' a different flood.

---

[90] 2 Peter 3:5-7 (KJV): ***For this they willingly are ignorant of, that by the word of God the heavens were of old, and the earth standing out of the water and in the water: Whereby the world that then was, being overflowed with water, perished: But the heavens and the earth, which are now, by the same word are kept in store, reserved unto fire against the day of judgment and perdition of ungodly men.***

[91] Genesis 9:1 (KJV): ***And God blessed Noah and his sons, and said unto them, Be fruitful, and multiply, and replenish*** *('mala')* ***the earth.***

[92] Genesis 1:28 (KJV): ***And God blessed them, and God said unto them, Be fruitful, and multiply, and replenish*** *('mala')* ***the earth...***

[93] Psalm 104:30 (KJV): ***Thou sendest forth thy spirit, they are created: and thou renewest the face of the earth.***

[94] Psalm 104:6-7 (KJV): ***Thou coveredst it with the deep as with a garment: the waters stood above the mountains. At thy rebuke they fled; at the voice of thy thunder they hasted away.***

**Job 9:7**[95] describes God 'switching off' ('***and sealeth up the stars***') the Sun and stars. When did that happen if not in a 'gap'?

**Isaiah 40:26, Isaiah 45:12 and Isaiah 48:13:** it is my position that the '***he calleth them all by names by the greatness of his might***', '***all their host have I commanded***', and '***when I call unto them, they stand up together***' all also refer to God switching the stars on and off – off prior to Genesis 1:2 and on again in Day 4 (in an Electric Universe model which we will come to).

**Job 37:10**[96] describes God breathing on waters to form ice. When did that happen?

**Isaiah 14:13**[97]describes Satan saying he will ascend to Heaven (before his fall), and in the Bible the only place you ascend up from is Earth – Earth must have been present at Satan's fall[RDB]; yet he had already fallen to tempt Adam; so Satan must have fallen either during creation week or before then, and Earth must have been there for the latter.

**The KRM:** most notably of all, the KRM shows repeated metaphorising and comparisons with a previous world – everything is a metaphor.

There are more examples in the KRM book, but this is enough for here. I suggest the Christian-creation debate is over.

**Gap opposition**

A question that is always asked is: 'Why did God not mention the gap?' Well, the examples above say He did, but you have to remember that the whole of Genesis 1 and 2 carry the KRM; and there is no gap of a

---

[95] Job 9:7 (KJV): ***Which commandeth the sun, and it riseth not; and sealeth up the stars.***

[96] Job 37:10 (KJV): ***By the breath of God frost is given: and the breadth of the waters is straitened.***

[97] Isaiah 14:13 (KJV): ***For thou hast said in thine heart, I will ascend into heaven,***

million years or so in the salvation process. So, He cannot include the gap, as it would invalidate the KRM – every verse and word is in the KRM. Likewise, God describes stretching out the universe during His creation, and He describes it at least 14 times in the Bible – this is important by its number of mentions; yet again He also omits this from Genesis 1 and 2, and I suggest again that this is due to it not having a parallel in the salvation process (and with it happening many years prior to the Genesis six-day account).

The Bible says it was through 'Adam that sin entered the world'[98] (*kosmos*) and Gap Theory contradicts this? Not really: they are different worlds (*kosmoses*) – you could just as well say the New World to come also has sin from Adam in it – but that too is a different world (*kosmos*). Sin and death entered the first world (*kosmos*) through Satan, and this too parallels Man's spirit (*kosmos*).

How can God declare a creation good[99] if there are 'dead bones' and Satan in it? He is declaring His new work good – the aspects He has just made ('*everything he had made*'). He also declares Man will be '*without spot or wrinkle*' in Ephesians 5:27 (without sin and living in the flesh) – so also very good too within a fallen world.

My position is I am not 'rubbishing' other creation paradigms, as I will show that young Earth creationists, day-age creationists, old Earth creationists and even theistic evolutionists are all 'partly' supported by this new interpretation of scripture.

Now let's get down to detail…

---

[98] Romans 5:12 (KJV): ***Wherefore as by one man sin entered the world, and death by sin, and death passed upon all men, for all have sinned.***

[99] Genesis 1:31 (KJV): ***And God saw every thing that he had made, and, behold, it was very good. And the evening and the morning were the sixth day.***

# Chapter 16

# A Whale of a Difference

## *An explosion of understanding*

**A kind of proof of the KRM**

There is much dispute and discussion amongst creationists about the selective use of *'kind'* (*'min'*) in Genesis 1 – it is used ten times there. As you may expect by now, I have a completely different take on it and it will reveal much. Within the KRM *'kind'* means the family of God and Man's recognition of it. *'Kind'* is used or omitted for 18 different life forms, and it is the change in its use from the before- to after-salvation conditions that reveals its intended meaning. (The realisation of this was my eureka moment in discovering the KRM: I knew I was onto something from this time on. Throughout the KRM all 18 uses or omissions of *'kind'* are predictable by the KRM – that's about a quarter-of-a-million-to-one chance of being right randomly.) Below, life classed as a 'kind' is in **bold:**

Day 3: Before – grass; herb; **tree** (*'upon the earth'*)
Day 3: After – grass; **herbs; tree**

Day 5: Before – moving creature; fowl
Day 5: After – **whales; living creature; fowl**

Day 6: Before – **living creature:** cattle; creeping thing; **beast of the earth**
Day 6: After – <u>**the**</u> **beast of the earth; cattle; creeping thing** (*'upon the earth'*)

'*Upon the earth*' is defining an interaction with Man's spirit (earth), so a 'tree' or 'creeping thing' may not be recognisable to Man, but 'upon the earth' they are.

Before salvation, in Day 3, unsaved men (grass) are not in the family of God; nor do they recognise who Jesus Christ is (herb); but **all** do hear the voice of the Holy Spirit (**tree**) drawing them in. In Day 5, unsaved Man does not recognise the Holy Spirit (moving creature) or the gifts of the spirit (fowl). And in Day 6, Man does not recognise the indwellings of Man in Christ and Christ in man (cattle) or the Holy Spirit (creeping thing) but does recognise himself – also from God and of the family of God (**living creature**) and his own abilities (**beast of the earth**).

After salvation, in Day 3, unsaved men (grass) are still not in the family of God, but now saved men (**herbs**) are, and the Holy Spirit (**tree**) is no longer '*upon the earth*' but now in Man and so is still recognised. In Day 5, saved Man now recognises the indwelling Christ and Holy Spirit (**whales and living creature**) and the gifts of the spirit (**fowl**). In Day 6, saved Man now recognises his new body-of-Christ-like personality (**<u>the</u> beast)** and the indwellings (**cattle**) and the Holy Spirit (**creeping thing**).

Every single detail in these before-and-after verses has meaning. I know of no other explanation for the selective use of '*kind*'.

God uses general terms for animal life in Genesis like '*beast of the earth*' or '*fowl of the air*', but not always – He also uses, specifically, '*whales*' and '*cattle*', and these are both shown in the KRM to represent the indwelling of Christ and the Holy Spirit. If I consider the nature and design of whales with their metaphoric meaning in a biblical context, and if I couple this with the knowledge of the KRM... then **KABOOM BABY...**

(The following couple of pages are from the KRM book.)

## Zoological considerations

Whales are mammals adapted to a different environment. Mammals that breathe air and are warm- and red-blooded. Mammals that spout water. Mammals with varied and complicated communication systems. Mammals that alloparent. Mammals with complex working-together and synergistic skills. Mammals showing a geological record of an organism abruptly and significantly modified enough to be termed *'created'*: a warm-blooded, red-blooded, air-breathing mammal living in waters, **in the same way that Jesus became a warm-blooded, red-blooded, air-breathing man (mammal) so that He could live in Man – in His spiritual waters.**

Fascinating! A creation of whales just to 'type' the indwellings of Jesus and the Holy Spirit and also to 'type' some of the functions of the *'body of Christ'*? I believe there is huge significance to animal types, size, population numbers and distribution found both in the fossil record and even with animals living in the present day.

**I suggest there is a four-fold nature to Genesis and the KRM:**

1. **The first plain-text story of the remake of life on Earth, with a plain-text description of the previous world revealed by the phrases '*it was so*' and '*brought forth*'.**

2. **The second narrative prophetically reveals the roles of each member of the Trinity of God in Man's redemption, and also the interaction to achieve this is revealed by a series of 'typings' and a very precise and consistent use of specific words.**

3. **The third aspect covers both narratives in that the conditions and life (possibly including all fauna) in the first world will parallel Adam's pre-fallen spirit. The conditions and life (possibly including all fauna) on Earth after the six-day making account will parallel all redeemed men's spirit, with the same 'typing' throughout. The first world will reflect Jesus as God and the**

**second, Jesus as a man: two differently 'themed' worlds with biological similarity.**

4. **I suggest, the KRM is 'in-depth' reality: In effect, the description and actions of the Holy Spirit and Jesus Christ are symbolised by animal life in the KRM and the real world. God has designed and made animal life (both anatomically and in nature) to reflect this – in other words, God makes real life reflect the biblical metaphor which at the same time reflects Jesus, the Holy Spirit and mankind.**

**We are living in a 'living metaphor'. Incredible!**

## More bloody details

There is so much more that can be deduced from the use of '*whales*' in the KRM once we know they represent the indwelling Jesus Christ in Man (and once the use of '*kind*' is understood). '*Whales*' may be considered to be '*beast of the earth*' modified to live in the sea. This can be interpreted as God's (Jesus') own work; His own '*beast of the earth*' made these '*whales*'. The '*whales*' are mammals to reflect Jesus as a man, but their extremely large size, I suggest, reflects Jesus as God. The divine are recognised by stature.

In the first world, the fossil record shows there were huge sea reptiles, also a changed version of land animals that have been modified to live in the sea. This seems to be a way of defining a change that is needed to indwell Man, with the sea reptiles again representing the indwellings, but this time in Adam. (**Huge prehistoric sea reptiles testify to Christ!**) Perhaps the nature of whales and huge sea reptiles symbolises a presence in both the 'dry' part of Man's spirit and '*waters*' part of Man's spirit. Adam's spiritual gifts are associated with reptiles and born-again Man's with mammals (reptiles dominate the first world and the warm-blooded animals the second). Why make reptiles and then mammals? What about the rest of the fossil record?

By the use of '*qara*' we found that the '*beast of the earth*' represent the spiritual gifts, traits, identity and character that God gives to Man and how He gives them. I suggest God made animals to anatomically and physiologically reflect the nature of these gifts, and included within the parallel is how these gifts are to be nurtured. Reptiles and mammals have specific differences and similarities, and I believe God made these two animal classes to testify to Jesus as God and Jesus as a man. The main differences between mammals and reptiles are, respectively: being warm vs cold-blooded; teeth and jaw being more vs less refined; young born vs eggs laid; production of milk for and nurturing of young vs no production of milk and no nurturing of young; a four-chambered heart vs a three-chambered heart; no growth at adulthood vs continual life-long growth; skin vs scales; and generally more refinement, adaptation and specialisation in mammals.

***'God walking in the garden in the cool*** *('ruah')* ***of the day*** *(God's light)***'** of Genesis 3:8 is, I propose, an allusion to the presence of God in Eden. Reptiles are exothermic and need the environment for heat; and in parallel, Adam's '*beast*', I suggest, needed the presence of God to fully function. In contrast, the indwellings are sufficient for born-again Man, and the mammals that represent them are endothermic and do not need environmental heat in the same way as reptiles.

The Bible describes Man as needing to '*chew over the Bible*' to gain spiritual and biblical knowledge (Joshua 1:8[100]). Mammals, representing saved Man's spiritual gifts, have more refined jaws and teeth to '*chew over the Bible*'. Reptiles have less-defined jaws and less-adapted teeth, as Adam is given this knowledge in full and, I suggest, did not need to '*chew over the Bible*' (Genesis 1:26-28). This would

---

[100] Joshua 1:8 (KJV): ***This book of the law shall not depart out of thy mouth; but thou shalt meditate therein day and night, that thou mayest observe to do according to all that is written therein: for then thou shalt make thy way prosperous, and then thou shalt have good success.***

explain why some dinosaurs have apparently carnivorous teeth when they were originally, according to the Bible, herbivores. In effect, they are not herbivorous or carnivorous teeth – just less refined and more 'versatile'. As a dentist I can say that they could chew fine with them either way, by the way! (**My dentition and TMJ testify to Christ!**) The presence of a deciduous dentition may also be representative.

The Bible describes the Word of God as the *'milk of the word'*, needed for salvation and growth after that (1 Peter 2:2[101] and 1 Corinthians 3:2[102]). Mammals produce milk for their young and care for them and, in parallel, so do the gifts they represent in born-again Man; he needs the *'milk of the word'* for his spiritual gifts. Adam is given these gifts in full and, in parallel, reptiles lay eggs and do not produce milk or nurture their young.

The *'heart'* is likened to the spirit and soul of Man throughout the Bible. Adam's 'base spirit' and *'beast of the earth'* are from God the Father, and I will show (by the presence of a water vapour shield in the first world) how it is 'augmented' by the indwelling Holy Spirit and the Word (Jesus Christ); this totals three inputs – three 'lights'. Reptiles have three chambers in their heart (only one atrium) to symbolise this. Man is born again, and his original *'beast of the earth'* and *'living creature'* are from God the Father and this time augmented and changed by the Holy Spirit, Jesus Christ as God and also the Word of God (or perhaps more precisely the Word of God carrying the Gospel of Christ) – totalling four inputs – four 'lights'. Mammals have four chambers in their heart to symbolise this. (**My atria testify to Christ!**)

Reptiles continually grow, and Adam was not meant to die; his spiritual gifts were to be given in continuity. Animals stop growing,

---

[101] 1 Peter 2:2 (KJV): ***As newborn babes, desire the sincere milk of the word, that ye may grow thereby:***

[102] 1 Corinthians 3:2 (KJV): ***I have fed you with milk, and not with meat: for hitherto ye were not able to bear it, neither yet now are ye able.***

Man was cursed to die, and his spiritual gifts given on Earth are limited by time.

Reptiles have scales as a covering, while mammals have skin and fur or hair – I think this could also be relevant when paralleling the reveal of a sinful nature (Genesis 2:25[103]). *'Not having spot or wrinkle'* was Paul's way of describing holiness in Man (Ephesians 5:27[104]), and God made Man coats of skin to *'cover'* their nakedness (Genesis 3:21[105]). This would seem to fit with Man's *'beast of the earth'* being represented by mammals whose fur covers their skin, matching saved men's Christ-like new *'beast of the earth'*.

Adam's gifts belong to Adam, but Man's gifts are now in Christ's image and within the body of Christ.

I showed in Chapter 10 of the KRM book how the term *'kingdom of heaven'* is a *'likeness'* statement – a location or 'atmosphere'[106] where God's will and purpose are carried out; thus all saved men's Christ-like *'beast of the earth'* are in the *'kingdom of heaven'*. We can see in the New Testament parables (Matthew 13) what the *'kingdom of heaven'* is like, and we would expect to see these properties in Man's *'beast of the earth'*. These parables tell us Man's *'beast'* will be *'fruitful'* (*a hundredfold, sixtyfold or thirtyfold*); will be like a herb spreading, branching out and growing into a tree; and also will grow as yeast makes bread rise. Jesus also said He would give Man abundant life

[103] Genesis 2:25 (KJV): ***And they were both naked, the man and his wife, and were not ashamed.***

[104] Ephesians 5:27 (KJV): ***That he might present it to himself a glorious church, not having spot, or wrinkle, or any such thing; but that it should be holy and without blemish.***

[105] Genesis 3:21 (KIV): ***Unto Adam also and to his wife did the LORD God make coats of skins, and clothed them.***

[106] I first heard the use of the term 'atmosphere' of God by Robin D Bullock and particularly like it and its link with the term 'kingdom of Heaven'.

(John 10:10[107]) and likened Himself to a vine, saved men to branches, with these branches then producing fruit (John 15:1). With this in mind, I think the term '*be fruitful and multiply*' as an instruction to Man in Genesis 1:28 may also be generally misunderstood. '*Multiply*' is clearly God's general instruction and blessing to procreate in large numbers. It is generally thought that '*be fruitful*' is another re-emphasis of the same, but, I suggest, God is not repeating Himself. '*Be fruitful*' is, in the literal, a command for Adam to be Christ- and Holy Spirit-like, and productive; in the KRM it is for saved Man to be the same. Therefore, in the '*beast of the earth*' parallel, there must be some anatomy or nature that represents this fruitful nature in animals.

Then we come to mammals that show more adaptation, diversification and specialisation. I propose that this represents the Christ-like personalisation of saved men's '*beast of the earth*' reflecting the properties of the '*kingdom of heaven*'. Another reason for mammals being abundant, adaptive and diverse is to cope with severe weather, seasons, changing conditions and catastrophes (all of which may represent sin). Adam's greenhouse world (we will come to this in Chapter 17) represents his sin-free status, and needs no such refined adaptations.

Genesis 1:28[108] and Genesis 1:22-23[109] may even be considered as 'diversification of species' statements or more properly 'diversification of species within *kinds*' statements. For very precise reasons, Verse 28 leaves out the term '*beast of the earth*' and instead defines what these

---

[107] John 10:10 (KJV): ***The thief cometh not, but for to steal, and to kill, and to destroy: I am come that they might have life, and that they might have it more abundantly.***

[108] Genesis 1:28 (KJV): ***And God blessed them, and God said unto them, Be fruitful, and multiply, and replenish the earth, and subdue it...***

[109] Genesis 1:22 (KJV): ***And God blessed them, saying, Be fruitful, and multiply, and fill the waters in the seas, and let fowl multiply in the earth.***

'*beast of the earth*' will be like; they will be '***blessed***' (gain something or have something dormant awakened), '***be fruitful***' (branch off and produce new fruit, increase, grow and become abundant), '***multiply***' (increase in number), '***replenish***' (restock and 'fill' something originally stocked) and '***subdue it***' (overcome whatever challenges the world presents).

This almost appears to be a call for animals to diversify to cope with any changes in the world. I personally find adaptation and diversification astonishingly beautiful and I would expect anything that parallels Christ to be astonishingly beautiful. (**Darwin's '*fruitful*', multiplying, replenishing and subduing Galapagos finches actually testify to the body of Christ!**) My position is that animal life is a dynamic and will reflect three different spiritual statuses: Adam, the Abrahamic Covenant and the better covenant of Christ. They will all command significant changes to life on Earth to reflect these statuses. It may be that animals with inherent genetic potential epigenetically change to metaphorically represent Man between certain times on Earth. The same holds true for fish and birds to be '*blessed, fruitful and multiply*' (Genesis 1:22). I believe the '*blessing*' is the second-law-of-thermodynamics-busting- and God-ordained process that the secular world interprets as evolution by 'random mutations'.

I think there will be other significant reptile–mammal differences, especially with regard to blood constituents, biochemistry and maybe DNA. There are other important parallels too.

Spiritual gifts that are owned by Man are 'typed' by warm-blooded, red-blooded, air-breathing animals ('*whales*', '*fowl*', '*beast*' and '*cattle*'). Spiritual gifts that are acting in Man and the Holy Spirit together (or that remain, at least, in part controlled by the Holy Spirit) are represented by '*creeping things creepeth*', '*moving/living creature moveth*') and are non-warm-blooded and non-red-blooded (smaller sea creatures, invertebrates and insects).

As well as Man being a 'blood man', Man and God are under a 'blood covenant': '*Blood is the life*' (*Deuteronomy 12:23*[110]). This results in other different indwelling parallels, with the specific nature of some blood-carrying animals ('*cattle*') as well as '*whales*' being used to emphasise and parallel Christ's role.

This in turn may shed further light on possible parallels with the size and distribution (and type) of reptiles, insects, birds, all animals and especially mammals in the **pre-KT and post-KT extinction-level fossil record (where I place the gap in Gap Theory).**

Again, it *may* be that air-breathing represents the notion that the gift is 'air-breathed' into born-again Man from the Holy Spirit (on Earth). '*Fish of the sea*' are not 'air-breathing', and so are not directly air-breathed from the Holy Spirit – but they are red-blooded, as they are Man's own, '*living waters*' that have just been enhanced by the Holy Spirit. The '*living water*' comes initially from God, and so it is '*the gift of God*' (John 4:10[111]). Smaller sea creatures (some of the '*living creature[s]*') and insects ('*creeping thing[s]*') are also not air-breathing or red-blooded, and so remain 'attached' to the Holy Spirit. (Breathing is classed as the ability to inhale and exhale through lungs.)

With all this in mind, we should expect to see not only the appearance of modern-day whales (or a recognisable predecessor) being associated with the KT extinction level, but also specific changes in other life forms. As well as that, we should see another 'further refinement or the result of more *fruitfulness*', at another specific time in history – I will show how the Abrahamic Covenant will be symbolised by life at the time of the Genesis six-day making account, and life after

---

[110] Deuteronomy 12:23 (KJV): ***Only be sure that thou eat not the blood: for the blood is the life; and thou mayest not eat the life with the flesh.***

[111] John 4:10 (KJV): ***Jesus answered and said unto her, If thou knewest the gift of God, and who it is that saith to thee, Give me to drink; thou wouldest have asked of him, and he would have given thee living water.***

Noah's flood will represent the better covenant of Christ. Thus we would expect to see in the fossil record that confirms this:

- A sudden increase in mammalian birds and bird types (and flightless birds to parallel the use of *'fowl that may fly'* in Genesis 1:20) with some wholly new designs (*'created'*) post-KT extinction to correspond to winged fowl – i.e. Holy Spirit gifts. But, as Jesus is a man now, these birds are warm-blooded and red-blooded; according to *'his kind'* – Christ (as a man).

- Larger sea creatures pre-KT extinction to correspond to Adam's greater gifts in his spiritual waters (this may also be to reflect a 'divine' element to these gifts symbolising Jesus as God), and also precursors to whales that will be used to create whales. (These may even be land animals.) An abundance of sea creatures, both blood- and non-blood-containing, appearing post-KT extinction level to correspond to the abundance of spiritual gifts (i.e. *'abundant living creature[s]'*). This leads to a sea teeming with life.

- Because of the use of the word *'create'*, we should see whole new designs of many other sea creatures (*'moving creature[s]'*), but with at least some biological similarity between pre- and post-KT extinction, as all are based on what the waters *'brought forth'*.

- A whole range of copies of sea creatures from pre- to post-KT extinction that were just brought forth with minimal differences.

And with Day 6 – *'beast, cattle and creeping thing'* – we should expect:

- Larger land animals and larger reptiles pre-KT extinction (dinosaurs) to correspond to Adam's greater gifts and to recognise a divine element to them too. These do not 'have to be' mammals according to Jesus as a man's *'kind'*, but are still red-blooded, as they parallel Adam's gifts, reptiles and mammals both being *'beast of the earth'*. **(T. rex testifies to Christ!**) A bigger overlap of *'beast of the earth'* and *'fish of the sea'* with more amphibians and sea-dwelling *'beast* *of the earth'* pre-KT extinction. This is because Genesis 1:9-10[112] describes the first world as having *'waters'* and *'dry ground'* (*'it was so'*) and our second world as having *'seas'* and *'earth'* (*'it was good'*). (Both have one land mass.) *'Seas'* denotes a demarcated body of water and *'waters'* the substance. Likewise for *'earth'* and *'dry ground'*. (Proverbs 8:29[113] adds detail in that *'seas'* are bounded.) Thus, the *'waters'* show greater intermingling with land (vast marshes and floating mats of land) in the first world, and, by extension, the same overlap with animal and aquatic life. The whole intermingling of waters and land, reflecting the indwelt Adam having a greater and more dynamic relationship with the Word (Jesus Christ) and the Holy Spirit. All of their spirit and spiritual waters are more closely entwined. I will show how smaller creatures revealed in the Cambrian, and plants overall, may support this.

- A whole range of copies of animals from pre- to post-KT extinction that were just brought forth and remade with minimal differences.
- A change after the KT extinction – as beast are made *'after his kind'*, with the *'his'* being Jesus as a man – meaning significantly more mammals and even primates to reflect this. Are primates just an animal as close to *'his kind'* as an animal can be? (I emphasise that this is not *'in his image'* – this is referencing Jesus as a man, not a spirit.)
- The appearance of domestic animals (*'cattle'*) now as mammals

---

112 Genesis 1:9-10 (KJV): ***And God said, Let the waters under the heaven be gathered together unto one place, and let the dry land appear: and it was so. And God called the dry land Earth; and the gathering together of the waters called he Seas: and God saw that it was good.***

113 Proverbs 8:29 (KJV): ***When he gave to the sea his decree, that the waters should not pass his commandment: when he appointed the foundations of the earth:***

according to Jesus as a Man's kind post-KT extinction. The appearance of animals to be used to represent Jesus Himself – lions or their predecessor kind.

- An abundance (diverse and large numbers) of non-red-blooded animals – both aquatic and terrestrial – to parallel the use of *'in abundance'* in Genesis 1:20-21, and possibly to also parallel the Holy Spirit acting on and approaching all men many times. (The *'creeping thing'* of Verse 24 includes within it the subset of *'abundance'* of Verses 20 and 21.) The increase in numbers and wide distribution of these animals may also parallel the Holy Spirit's omnipresence. Thus an increase in insect and smaller sea creature species would be expected, and this is confirmed today, especially with regard to insects. (Did you know that only approximately 250,000 species have been identified in the entire fossil record, and that presently on Earth there are about 2 million species already identified – about 1 million of these being insects – 'fruitfulness' on display.) Also, there are about 250,000 marine species alone identified, with some speculating that this is only about one tenth of the actual amount. Armed with this knowledge, I think I could probably have zinged J. B. S. Haldane's (in)famous quote had I been there[114] – *'to testify of Christ's and the Holy Spirit's work, obvs, JB!'*
- I believe the first world had a very misty atmosphere[KK], this being essential for dinosaurs to survive (rather like alligators of today, whose skins fatally crack open unless continually moistened). Perhaps this also parallels the huge gifts of Adam needing the Holy Spirit to be enabled (He is metaphorised as mist in Genesis 2:6) – this will 'blur' the land-water barrier; it is consistent with Adam requiring this presence. (I believe two thirds of all Earth's water was

[114] It may be apocryphal, but Haldane was supposed to have sarcastically described a creator God as being 'inordinately fond of beetles', there being so many species of them – about 400,000 at the last count.

present in the water vapour shield[KK] – this parallels the Holy Spirit and Jesus Christ residing in Heaven.)

- **Please note** that although the act of *'bringing forth'* implies a similarity between the worlds, it seems that there are very significant similarities to God's 'nature' in nature only when the life *'brought forth'* is described as a *'kind'*.
- In consistency, it would be expected that plants would follow the same pattern (Genesis 1:11-12). Huge trees reflecting Adam's bigger gifts and a divine element to them in the first world, and smaller but more refined, proliferative and diverse trees and herbs in the second. The mustard seed parable (Mark 4:32) may also refer to this, and it may make the roosting *'birds'* of this parable a reference to gifts of the spirit or angels or both. (The herb becoming a tree may even be a *'likeness'* becoming an *'image'*.)

**All of these observations are confirmed by the fossil record. All of these predict the fossil record. The fossil record that has been used by atheists for years to disprove the existence of God may actually be testifying to Christ. Well, well, who would have thought that?** Colossians 1:16 tells us that '... ***all things were made by him and for him***', the *'him'* being Jesus Christ – we should expect life on Earth throughout the eras to reflect this? *'Suggest name change from KT extinction event to **JC distinction event!**'*

Interestingly, the KT extinction event marks the beginning of the last of the geological eras – the Cenozoic era which we are currently still in. The literal meaning of 'Cenozoic' is 'new life'. This name fits well with Gap Theory, which also defines wholly 'new life' beginning on a judged Earth at this point in time. It also works well with the KRM, as 'new life' represents aspects of the salvation process by which the Triune God gives Man a 'new life' – the born-again Man is 'mentioned' in geological era terminology.

For these reasons, to pay respect to the term 'young Earth creationism' and to show that the paradigm I am describing is not really that different, I have chosen the new term **'young Cenozoic creationism'**, which will again be described in much more detail in Chapter 17. I also like the term **'born-again creationism'**, as this simultaneously links the process to Man's salvation, and so combining the two we have:

**'Young Cenozoic (born-again) creationism'.**

## Leading lights and radiometric dating

We have already discussed the subtle wording of Day 4 and how 'lights' can have multiple metaphoric meanings, but there is more to the plain-text meaning of Day 4, especially when combined with Genesis Day 1 – and these, I believe, may give a reason as to why radiometric dating gives such a huge age to planet Earth.

**Day 4** (Genesis 1:14-16)

> 'Pre-salvation – **Genesis 1:14-15** (KJV): ***And God said, Let there be lights in the firmament of the heaven to divide the day from the night; and let them be for signs, and for seasons, and for days, and years***. (Jesus shining light on the whole world.)
>
> ***And let them be for lights in the firmament of the heaven to give light upon the earth: and it was so.*** (This is now giving light upon the '*earth*' and not just '*in the firmament*' – the spirit of individual men.)
>
> 'At salvation – ***Genesis 1:16-18*** *(KJV):* ***And God made two great lights...***

There is great subtlety in this text. Usually in Genesis 1, there is a one-sentence '*it was so*' statement followed by a one-sentence '*it was good*' statement, with a long-term-after statement often following that. However, in Day 4 there are two statements that precede the '*it was so*' proclamation. The 'it was so' has three meanings:

- a literal description of what God will be making for our world
- a description of God's interaction with Man pre-salvation – Jesus will be the light of the whole world (Verse 14) and be presented as the light of life to individual men (Verse 15)
- a description of the first-world conditions.

The '***And let them be for lights in the firmament of the heaven to give light upon the earth***' seems repetitive in the literal meaning, and almost not needed, as it is very similar to the previous verse. It has significant meaning in the KRM, but in the first-world description it raises the possibility that the '*it was so*' may apply only to Verse 15 and not Verse 14 (other '*it was so*' declarations apply only to their preceding sentence). Thus the '*it was so*' could be describing a world lit by 'lights' only and not necessarily lit by the Sun and Moon – a world lit by stars, and with a Sun and Moon much less 'noticeable' or even not noticeable at all. I believe this is the reason for this additional sentence (please note that no Sun or Moon will result in no tides).

**An astronomical aside.** Thus, suggesting that stars ('*lights*') were visible in what I will show to be a constant daytime in this first world (this will be more understandable with the Earth being lit from 'within', and with the presence of a glass-like water vapour shield with no sunlight to burn out the light of the stars – I will show how the Bible describes all of these conditions). Please note: present-day stars can sometimes be seen in daylight.

**A dreamy aside.** I believe that dreams are God's way of occupying the spirit which does not sleep. During dreaming, when we are 'in the spirit', we lose all track of time. As Adam's pre-fallen spirit is highly spiritually blessed – 'in the spirit' – and as the first creation parallels Adam, no discernible way to measure passing time with days and years seems to fit.

**Day 1: Light leaders**

> God says 'I am': ***Genesis 1:3-4** (KJV)**: And God said, Let there be light: and there was light.***
>
> (Post-salvation – He does not change): ***And God saw the light, that it was good: and God divided the light from the darkness.***

When God declared '***Let there be light***', seemingly immediately light then '*was*' (this light is from the spiritual dimension, as nothing is worked on or made to produce light in the physical realm yet). This immediate response seems to be almost an instant and 'always was' version of '*it was so*'. **Recap:** God the Father declares light to be, so light 'types' God the Father. This phrase '*there was light*' is, I suggest, alluding to the immutable and infinite God not changing; it is a kind of '*I am*' statement for the KRM and an '*I am*' statement for His Glory Light in the literal first creation too. **'Light Was'.**

So, by considering this as a modified '*it was so*' statement, this '*light*' must also be applying to the first creation – God's Glory must have lit the first creation, and this will parallel God's Glory lighting Adam's spirit and '(it) *was* (so)' (in the first creation).

From this, together with Day 3 stating there was no Sun or Moon, and with the first world being a parallel for Adam's pre-fallen spirit (which had no darkness in it), **we can deduce that the first world was lit by the Glory of God with no night-time or darkness.**

In the KRM, God does indeed shine a light for Man's path before salvation, and '*it was so*' applies to that too. '*It was good*' also applies to both the literal and the KRM – but there are two grades of light here. God lighting up the first world is actually paralleling light from **within** Adam's spirit – an indwelt Adam. The light comes from within the first world – a trans-dimensional statement. The light of Genesis Day 1 shines light **on** the creation and shines a light for Man's path – not from 'within'. Three days later, the Sun and Moon are made by Jesus' work and they light the world – three days after Jesus is in the grave, He lights saved Man's spirit.

**Please note:** All life forms described as '*kinds*' has real-life metaphors – specific parallels in real life to confirm the nature of what God has given. I suggest that God is using '*kind*' to show the importance of this. In a way, this life is, metaphorically, in the family of God too. The first world (*'trees', 'the living creature' and 'beast of the earth'*) and the second remake (*'herbs', 'trees', 'whales', 'living creature', 'fowl', 'beast of the earth', 'cattle' and 'creeping thing'*) will all have real life as their metaphor, as all are described as '*kinds*' – all 'kinds' will have a specific nature or design to reflect this.

Now let's see the history of absolutely everything just from Genesis...

# Chapter 17

# The Long Life – Long Gap – *Genesis* Six Days Model

## *Young Cenozoic (born-again) creationism*

*Every fossil layer and the four eras explained*

(This chapter is a close copy of Chapter 23 in the KRM book.)

**Stretching**

God decided it was time to create the physical realm.

The Word (the Lord Jesus Christ) travelled within God **(*'the word was with God and the word was God'*)** to a location in the spiritual realm, and then stepped out of God **(*'the same was in the beginning with God'*)** and stretched a vast scroll around part of (but still within) the spiritual realm (John 1:1-2[115]).

The Word continued to stretch out further parts of this scroll. He joined the 'ends' together after wrapping them around a part of the spiritual realm but with the whole expanding scroll still contained within the spiritual realm. The 'spokes' of this scroll are vast and expanding at the speed of light and almost beyond comprehension.

There are about 14 places where the Bible describes God stretching

---

[115] John 1:1-2 (KJV): ***In the beginning was the Word, and the Word was with God and the Word was God. The same was in the beginning with God.***

out the universe (Isaiah 45:12[116]), this high number of mentions indicates the significance God gives this process.

**God puts the '*mayim*'** (water) **in the '*samayim*'** (Heaven)

Water was involved in this stretching process (water was in the Word's hand at this beginning[KK]). I suggest, at this point, the Word places vast amounts of water in plasma form (hydrogen and oxygen), this plasma consisting of enormous sheets and filaments. These are placed throughout the entire Cosmos during the whole of this stretching-out and re-stretching-out process.

The Word then placed 'something else' in the physical realm on which to hang the Cosmos. I suggest this 'something' is the introduction and activation of zero-point energy and whatever the mysterious gravity derives from. A vast electrified fabric is thus created, comprising trillions times trillions times trillions of amps with enormous amounts of oxygen and hydrogen in plasma form. Fusion reactions fed by the resultant enormous currents quickly form all known elements in their current proportion.

**Living water too**

I believe that as well as water in the Word's hand, there was also a drop of living water from within the Father, which was incorporated into all of creation.[KK]

**About time**

Time begins, in the physical realm only, at this point. The physical realm (our Cosmos) is superseded by the spiritual realm, and so the physical realm can be folded, torn, punctured and rolled up because it is within

---

[116] Isaiah 45:12 (KJV): ***I have made the earth, and created man upon it: I, even my hands, have stretched out the heavens, and all their host have I commanded.***

and due to the presence of this 'superior' spiritual realm. Please note: the spiritual realm is as solid as our physical realm, not ghost-like and ethereal. The timed physical realm is different from the spiritual realm, where there is no time. (Revelation 8:1[117], Job 10:5 and 2 Peter 3:8 all hint at there being no time in Heaven.)

Firstly, God creates the planets and throws them out by hand one at a time; likewise, the stars and entire galaxies are thrown out by hand (Isaiah 40:12[118]). I suggest God throws out vast Birkeland currents of much, much more than 10 to the power 40 amps in strength, pulling in matter consisting of all elements now formed from the plasma filaments and sheets. This matter may be drawn from thousands of light years away. The plasma pinches and plasmoids, thus made, become all the galaxy shapes and sizes that are currently observed. The paired currents intermingle to form pinches which behave like gigantic whirlpools, swirling and drawing in the matter. This order of creation satisfies the order in Genesis 1:1. God throws these currents trillions at a time from a point in the spiritual realm, but they appear at many different locations throughout the physical realm (Hebrews 11:3[119]); when viewed these forming galaxies will appear to from quickly in a relatively short period of time.

God then creates the Earth, by speaking into a void, and things start to appear and spin.[KK] God creates ex-nihilo and, after some time, the Earth forms with one land mass surrounded by waters. God fills the

---

[117] Revelation 8:1 (KJV): ***And when he had opened the seventh seal, there was silence in heaven about the space of half an hour.***

[118] Isaiah 40:12 (KJV): ***Who hath measured the waters in the hollow of his hand, and meted out heaven with the span, and comprehended the dust of the earth in a measure, and weighed the mountains in scales, and the hills in a balance?***

[119] Hebrews 11:3 (KJV): ***Through faith we understand that the worlds were framed by the word of God, so that things which are seen were not made of things which do appear.***

waters with water in one hand (Amos 9:6[120] and Amos 5:8[121]); this is an indication that God can assume any size He likes.

**A speculative astrophysical aside** (based on the Electric Universe (EU) Theory and plasma cosmology)

There will be no Sun visible from planet Earth initially (it may look like just another star), and God's Glory will light the Earth. I speculate on three possibilities from a Gap Theory perspective, and bearing in mind that the use of the word *'asah'* in Genesis 1 probably indicates that the Sun and Moon already existed and were just 're-commissioned' in the six days of creation (Genesis 1:16).

I believe that the planets and moons have not moved from their 'position' since being originally placed there.[KK] In consistency with God's evident creative works, it would seem to be likely that a planet always orbits around a star.

God set (*'natan'*) the Sun and Moon in the firmament (Genesis 1:17), the word implying movement for one or both 'lights' from their original position. (Or Earth moving to them and the planets.)

Option 1: At this stage, some of today's planets (including Earth) orbit and spin around Saturn, which is a dwarf star at this time (there is evidence to support this, as described by Tom Findlay in *A Beginner's View of our Electric Universe*[122]; please note: his book is scientific and not theological).

---

[120] Amos 9:6 (KJV): ***It is he that buildeth his stories in the heaven, and hath founded his troop in the earth; he that calleth for the waters of the sea, and poureth them out upon the face of the earth: The LORD is his name.***

[121] Amos 5:8 (KJV): ... ***that calleth for the waters of the sea, and poureth them out upon the face of the earth: The LORD is his name:***

[122] T. Findlay, *A Beginner's View of our Electric Universe*, Guildford, UK: Grosvenor House Publishing, 2013.

Later, as described in Genesis 1, the Sun and its accompanying planets come into view from Earth by moving into close proximity with this dwarf star system. The Sun then takes over and dominates the dwarf star system and all of the planets. The dwarf star Saturn and its planets become planets around the newly arrived Sun. (This may, however, contradict 'the earth has not moved from its position'.)

Option 2: The Sun and Earth have always been in their current positions, but the Sun was very dull, with little light emitted. The EU models allow for stars to be 'switched down' to dark plasma mode. Here, from the beginning, Earth and all planets orbit the Sun. In Genesis 1 the Sun is re-ignited and moved a little relative to Earth, and the Moon is moved to its current location, the Moon having previously been far enough distant to not lead to a tidal sea. (It might still have been visible, though.)

Option 3: Earth is not in the solar system and is moved there in Genesis 1:17.

**Background to the Electric Universe model**

In the laboratory, all known galaxy formations can be reproduced by flowing two interacting electric currents (known as 'Birkeland currents') through plasma (the Big Banger's guilty secret!). In consistency with this, the remnants of currents can still be detected at right angles through the centre of galaxies.[123]

Over 99% of the observable Cosmos is plasma.

The standard model of Big Bang and cosmogony relies on the – so far unobserved – presence of vast amounts of dark matter, and on a personal level, I do not believe it.

---

[123] I highly recommend the Thunderbolts website, https://www.thunderbolts.info/wp/about/, as an introduction to the Electric Universe Theory. This is a non-religious website.

The EU model's weakness is its lack of an energy to start and to drive the drive whole process – well, I can answer that: because 'God meted out Heaven with the span' (Isaiah 40:12). Now back to the history of everything...

## A perfect harmony epoch begins, lit by the Glory of God

God creates a diverse range of plants, with huge trees hundreds of feet tall and flowers 30 feet high.[KK]

A vast number of sea creatures are created and placed in the sea.

God forms a transparent shield around the Earth, made of two layers with water in between (Job 37:18[124]) There is twice as much water enclosed within this 'water vapour shield' than is present on the Earth.[KK]

Water pours out from this shield and turns to mist on hitting the (warm) air, and the whole world is bathed in this mist under a peach-coloured sky.[KK] The mist comes from the water canopy downwards, but in the pre-fall second creation, mist comes from the ground upwards (Genesis 2:6[125]). Man could not survive in these misty conditions.[KK]

There is no rain, no thunder, no lightning and no real weather as we know it in this greenhouse world. The Moon is not in its current location, and the Sun is in dull, 'dark plasma' mode, not visible or as close as present, so by extension there are no tides. There are no meteorite impacts, no electrical discharge machining (EDM), no volcanic activity, no earthquakes – just a flat, calm sea (Job 9:8[126]). God creates the dinosaurs, some much bigger than already discovered and

---

[124] Job 37:18 (KJV): ***Hast thou with him spread out the sky, which is strong, and as a molten looking glass?***

[125] Genesis 2:6 (KJV): ***But there went up a mist from the earth, and watered the whole face of the ground.***

[126] Job 9:8 (KJV): ***Which alone spreadeth out the heavens, and treadeth upon the waves of the sea.***

some up to 90 feet tall.[KK] The order of plants, sea creatures then land animals is the same in both creation/making accounts. The distribution, size and type of animals and plants will satisfy a metaphor for Adam's spiritual gifts and reflect Jesus as God.

All live in perfect harmony, with no death of '*nephesh chayyah*' animals – animals with a spirit and soul (Psalm 104:30[127]).

There is no carnivorous activity by '*nephesh chayyah*' animals – they only eat plants (dinosaurs included), with less refined teeth, designed for very dense and fibrous fruit and vegetation. '*Nephesh chayyah*' animals are not part of any food chain, but non-'*nephesh chayyah*' animals may eat each other and be eaten. Death does not exist for these '*nephesh chayyah*' animals as defined by the spirit and soul leaving the body, leaving the possibility of the body and spirit and soul leaving the Earth together still open. Death, as defined in our modern terms, exists for non-'*nephesh chayyah*' animals and plants. There is no disease and no genetic abnormalities. '*Nephesh chayyah*' animals mature but do not age, and their reproduction is bounded.

The world remains lit by the Glory of God. God imparted this Glory just by touching the water vapour shield.[KK] God's Glory continues to shine from 'within'. This light permeates throughout the world and is different to the light of today, and there is no night-time and reduced shadowing – a magnificent creation under a peach sky, but still a little dimmer and a little 'lower' version of Heaven (Hebrews 8:5[128]). Stars are visible during the continual daytime due to the absence of the Sun (and its light-scattering properties), the presence of a 'highlighting' glass-like water vapour shield and light originating from 'within'.

---

[127] Psalm 104:30 (KJV): ***Thou sendest forth thy spirit, they are created: and thou renewest the face of the earth***.

[128] Hebrews 8:5 (KJV): ***Who serve unto the example and shadow of heavenly things...***

**Another speculative astrophysical aside of three lights:** God's Glory Light, pre- and post- the 'emergence-of-sin' light (based on an EU model and plasma cosmology).

More on the electric universe and plasma: The Electric Universe model describes plasma and electromagnetic forces as being the dominant matter and forces in shaping the Cosmos.

When sin enters the world, the whole of creation groans and not just part of it. ('Groan': a creaking sound made by an object under duress.)

> **Romans 8:22 (KJV): *For we know that the whole creation groaneth and travaileth in pain together until now.***

Please note: the 'whole creation groaneth' until 'now', so the 'groaning' may have stopped or may not be the 'whole' any more. (The timeframe for this 'now' is around the time of Christ; the 'groaning' is at the fall of Adam.) I suggest this means that when sin enters either world, everything in the creation groans: weather, organisms, DNA, zero-point energy, atomic energies and movements, and light (with light speed included).

**A variable speed of light:** This is a controversial subject, not least because a variable speed of light (VSL) would seem to contradict Einstein's theory, but more researchers are starting to suggest there was a quicker speed of light in ancient times, largely due to the 'horizon' problem. I speculate that the speed of God's Glory Light is extremely fast or even unmeasurably fast, and I suggest that its speed when unaffected by 'sin' and 'groaning' is much faster than the post-'groaning' light of today.

**Zero-point energy (ZPE):** ZPE is a vast ubiquitous energy that exists throughout the Cosmos and is thought by some scientists who specifically study this phenomenon to play a role in supplying the energy for atomic function and movements. Some scientists suggest that ZPE originates with the expansion of the Cosmos, and so increases as expansion goes on.

However, I tentatively suggest that ZPE in the initial creation – before sin enters the world – is more ordered and smooth, whereas, after the fall with sin and 'groaning', it becomes disordered and 'batters' subatomic particles more. The speed of light may, I also speculate, be slowed down due to this ZPE turbulence, rather than an increase in the ZPE strength due to cosmic expansion as some scientists claim. The ZPE pre-sin would be more like a smooth, swift-flowing river and not a choppy, turbulent one as it presents now. The light is like a canoe travelling down a rough- or smooth-flowing river. Instead of jostling the subatomic particles, ZPE would just impart more energy smoothly and with less magnitude of jiggling forces. As a result, 'in the beginning', the atomic velocities are also greater than today. These increased velocities lead to more frequent and more violent atomic 'collisions', and hence increased atomic decay and an appearance of age. It should be noted that the idea of a VSL due to a change in ZPE strength by an expanding Cosmos was first brought to my attention by Barry Setterfield on his website[129]; this is his idea and not mine. I just speculate on different drivers than Setterfield. I highly recommend both his website and his YouTube presentations. He is the only person in the entire creation debate that I would use the term 'genius' for, and I do not use the word lightly'; his 'out of the box' and data-driven thinking process encouraged me to do the same. It should also be noted that Setterfield may not support many of the views in this book.

**God's Glory Light:** I suggest God's Glory Light has unfathomable properties: it shines from within and there is no, or less, shadowing;

[129] B. Setterfield, Genesis Science Research, available at www.barrysetterfield.org/ (accessed 10 May 2023).

and it may even appear to 'permeate' matter due to its trans-dimensionality.

Hebrews 8:5 states *'Earth is a shadow of heavenly things'*, and I suggest that God's Glory 'dominates' the physical realm. I speculate that God's Glory Light supplies its own ZPE, and that all atoms subjected to it are energised more than under our present ZPE conditions, and electron emissions and absorption greatly increase under its influence. The remnants of God's 'eternal' Glory Light may leave a cosmic microwave background (CMB) behind.

**'Pre-groaning' or (pre-fallen Adam's light) in the second creation/remake:** I suggest that this light is not impinged on by disordered ZPE, so atomic decay is higher and the speed of light is faster than after the fall ('groaning'), for the same reasons as God's Glory Light but not to the same extent. The spiritual light shone **on** this world (not from within) but came from **within** in the first world paralleling Adams's indwelling presence then; but now it parallels God shining a path for Man to go to be redeemed.

**'Post-groaning' (post Adam's fall) light:** I suggest that light after the fall 'groans', which may well be because of the disordered ZPE. The light speed thus slows to today's rate. (That reduction may only just have bottomed out within the last 100 years.) Electron emissions and absorption are also at today's rate. I have a suspicion that it may speed up again in the Kingdom Age (that, I believe, has now arrived) to reflect God's Glory re-entering Earth (Habakkuk 2:14[130]).

**Electron excitability, emission and absorption**

Background: Electrons can absorb and emit light. Atoms may raise their energy level by absorbing light. Electrons have an 'urge' to return to their lowest energy levels and when they do this, they emit a photon

---

[130] Habakkuk 2:14 (KJV): ***For the earth shall be filled with the knowledge of the glory of the LORD, as the waters cover the sea.***

of light. A very specific amount of energy is required to make these electrons jump; too much or too little will not be effective. The light emitted is also specific to each element. I suggest God's Glory Light greatly enhances this process and is abundant in all of these energies.

To summarise, I suggest that in the early planet Earth, light speed is much quicker, all light throughout the Cosmos is blue-shifted, and the whole creation and notably ZPE and atomic movements are more ordered than at present. As atomic decay is quicker, this gives an illusion of greater age. Now back again to the history of everything...

## A Greenhouse World – the Precambrian Era

### This is God's perfect world

The greenhouse world has vast delicate ecosystems, all more demarcated than in the present world. The ecosystems are stable and well defined, with flora and fauna that are entrenched and very specifically adapted to their environment. Animal migrations are not happening, due to the abundant food available. The seasonal-changes driving of our changing ecosystems of today are not happening either. A slow flood catastrophe (originating from the sea) hitting this world could 'sample' these very demarcated ecosystems (multiple floods, and slowly rising sea levels likewise). A catastrophe that moves from sea inland could lay down strata containing fossils in layers, reflecting these clearly defined and different ecosystems with significantly different flora and fauna in each. This would give the illusion of organisms evolving from sea to land animals.

A different hydrological cycle is in effect, with water falling from the water canopy. Rivers and streams are rare and much reduced in size compared with the present day; they also flow at a reduced speed. Very little erosion occurs, so very little sedimentation and virtually no fossilisation take place except for some microbe fossilisation, which is also part of God's plan for land expansion and growth. There is no 'wasted' death to fossilise; all dead animals (non-'*nephesh chayyah*') are scavenged to base constituents both in sea and on land, and likewise plants. The world is teeming with life and with dense, lush vegetation.

There are no humans in this creation, but it is tended by many Watcher Angels, who are fleshy (carbon based) and build their own dwellings with their own domesticated animals (Daniel 4:13[131], 17, 23).

**God does not do 'monotony'**

A constantly changing creation seems to be God's plan; all creatures change within their '*kind*'. Genetically rich animals are the original creation, and they always retain their form or '*kind*' during this diversification process. There are no genetic mutations or genetic degradation in this world, just inbuilt genetic potential being realised. This world thrives for millions of years, and the Watcher Angels may occasionally leave their footprints alongside dinosaur tracks, which may also (very occasionally) become fossilised at the beginning of the soon-to-come catastrophic events as organisms flee. Animals also make footprints all the time, but none are preserved due to the lack of sedimentary fossilisation processes.

All organisms or their same '*kind*' predecessors that are found in the Archaeozoic, Palaeozoic and Mesozoic fossil records are present on the Earth at this initial point in creation. There has been only one creative event carried out by God at this stage.

There is a huge 'potential' for sedimentary fossilisation throughout the whole world, as there is no erosion by rain (unlike today), so potential sediments are not washed away by rain, fast-moving rivers, streams or tides.

---

[131] Daniel 4:13 (KJV): ***I saw in the visions of my head upon my bed, and, behold, a watcher and an holy one came down from heaven;***

**Vast floating mats:** Kurt Wise[132] theorised about the presence of floating mats of flora and fauna in the early Earth, likening them to the 'quaking' bogs of today. (It should be noted that Wise may disagree with many of the ideas in this book.) He describes how evolutionary theory for plants is supported by the plant fossil record through what he very aptly terms 'terrestriality'. Basically, greater terrestriality allows a plant to live further from standing water. Plant fossils show, chronologically, greater and greater terrestriality (meaning the plants lived further and further from open water). He also describes how hollow trees found within the fossil record have hollow root-like structures that could not have grown through soil but would have allowed the trees to float in water. He suggests that these floating mats caught in a flood could explain some aspects of the stratigraphic column.

I entirely agree with his insight and would like to extend this concept. I suggest these trees would not need to withstand strong weather either, because, in this 'greenhouse world' first creation, there was no such weather. However, this changed when the perfect ecosystems were damaged by Satan's fall and onslaught. I suggest that these floating mats were vast and surrounded the whole land mass, extending for hundreds of miles (maybe even the bulk of land was like this – remember that God's Glory Light can fully illuminate under these mats). I also suggest the sea was much shallower, extending

---

[132] K. P. Wise, 'The Pre-Flood Floating Forest: A Study in Palaeontological Pattern-Recognition', Proceedings of the International Conference on Creationism: Vol. 5, Article 31, 2003, available at https://.digitalcommons.cedarville.edu/icc_proceedings/vol5/iss1/31/ (accessed 10 May 2023).

K. P. Wise, 'Evidence that Noah's Flood Formed the Fossil Record – Dr. Kurt Wise (The Palaeontology of the Flood)', YouTube, 2018, available at https://www.youtube.com/watch?v=xCdt6LjvFvA (accessed 27 November 2025).

significantly further out than at present, and that it also was calm due to there being no weather or Sun-and-Moon-induced tides. There was only one third of the current amount of water on Earth at this time – two thirds being in the water vapour shield. Colonisation by single-cell organisms living under these mats, and subsequent fossilisation, is part of the process of this land mass expanding – again, God's Glory Light illuminates 'deeper' than we find today. This deeper illumination provides significant light under the floating mats for life to thrive, and also below that. This will provide a parallel with Adam's initial indwelling before the fall – a greater mingling of God and Man's '*living water*' and base spirit ('*dry land*') of Genesis Day 3.

The conditions under these mats can be compared to our present-day 'dark biosphere', where the lack of light and physical constraints mean they can be inhabited only by single-cell organisms. As God's Glory Light permeates further and comes from 'within', compared with our 'normal' light of today, so these microbes are situated much deeper than our present examples – they are the only organisms that can live there. As the land expands, these areas get thicker and also colonise downwards into the created terrain, leading some areas to grow up to several miles thick with single-cell microbes.

The Precambrian strata is laid down and is part of God's continually changing and growing creation process. The only fossilisation is of single-cell organisms along with their colonisation of deep terrain. Nothing else fossilises. There are no other processes that lead to organisms being buried, no mudslides, no extreme weather; any non-'*nephesh chayyah*' animal that dies is quickly scavenged away.

Millions of years pass and this perfect world is tended by Satan and his angels who are readying Earth for God's special new creation; this creation is to be made in the image of God in both His triune and trinity status.

***'Fiery the angels fell; deep thunder rolled about their shores'*** (A quote from Blade Runner's[133] adaptation of William Blake's poem *America a Prophecy*.)

The slow destruction of planet Earth is about to begin.

Satan is allowed access inside God, and to the Holy places situated there too. He witnesses the worship and the effect it has on God, and decides he wants it for himself. He travels around Heaven, canvassing support and displaying his beauty and talents. This takes some time; every angel is tempted and (shockingly) a third of angels succumb (Revelation 12:4[134], Ezekiel 28:16[135]). It is only worship angels who fall, and not all of them. Man is to replace them in eternity.[KK] (It still amazes me that beings with a full knowledge of God will rebel against him.)

A major part of this is Satan being enraged at discovering, and also feeling envy towards, Man's proposed creation with the vast gifts and attributes Man will have – he wants them for himself, and wants to have a fleshy vehicle to possess too.[RDB] Satan starts to destroy the Earth even before his judgement; Satan and his soon-to-be-fallen angel followers had some kind of governance role in the physical realm. The *'in his image'* (of Genesis 1:26) and *'like the most high'* (of Isaiah 12:14) are synonymous, this meaning Satan is wanting to be like Man (in design). If he cannot, he will destroy the Earth where Man will be homed. God knows what is happening (and may be testing the angels) but lets it proceed. At a certain

---

[133] R. Scott, Director, *Blade Runner* (film), The Ladd Company – Michael Deeley, 1982.

[134] Revelation 12:4 (KJV): ***And his tail drew the third part of the stars of heaven, and did cast them to the earth: and the dragon stood before the woman which was ready to be delivered, for to devour her child as soon as it was born.***

[135] Ezekiel 28:16 (KJV): ***By the multitude of thy merchandise they have filled the midst of thee with violence, and thou hast sinned: therefore I will cast thee as profane out of the mountain of God: and I will destroy thee, O covering cherub, from the midst of the stones of fire.***

point God reaches His tolerance limit and evicts Satan from Heaven *'like lightning'* (Isaiah 14:12-14[136] and Luke 10:18[137]).

The angels who fell were evicted likewise (by vastly charged arc plasma lightning bolts, perhaps?) consecutively, one at a time. This angelic eviction took 'some time'. The lightning-like event traverses the spiritual and physical realms, and lightning and thunder are seen and heard in both dimensions, for the first time on Earth.[KK] We can only speculate at the power of such a trans-dimensional phenomenon. The cargo-carrying lightning pierces the canopy, and it also pierces the sea and land, while depositing Satan and the fallen angels through the Earth or sea to some depth (not just onto the surface) and into Hell, which has been prepared as a domain for them (not a prison). It is not just lightning, but a large bulk arc plasma 'train' carrying a 'cargo' of Satan and his demons, one bolt for each one, consecutively.

There is no mass death and fossilisation in God's creation, so the appearance of mass fossilisation (the Cambrian) corresponds to this initial attack on creation from, and the knock-on effects of, this judgement (and also from an initial attack from Satan – see below).

One third of the angels fell. We do not know how many that entails, but we can guesstimate a minimum number. I believe that every person has at least one guardian angel and many have several.[KK] I also believe that it can take a thousand demons to have an effect on even one person.[RDB] With a world population of over 8 billion, this leads to a guesstimate that a third of all angels is certainly in the billions and, I

---

[136] Isaiah 14:12 (KJV): ***How art thou fallen from heaven, O Lucifer, son of the morning! how art thou cut down to the ground, which didst weaken the*** nations! ***For thou hast said in thine heart, I will ascend into heaven, I will exalt my throne above the stars of God: I will sit also upon the mount of the congregation, in the sides of the north:***

[137] Luke 10:18 (KJV): ***And he said unto them, I beheld Satan as lightning fall from heaven.***

suggest, likely trillions. God's 'aim' of the lightning bolts means there will be no death of '*nephesh chayyah*' animals by this judgement process or any follow-on disasters. Trillions of angels falling like lightning at great speed (speed of light) makes this likely to be a very violent event, and I suggest this contributes to the first worldwide catastrophic event on Earth – previously referred to as the misnomer: the 'Snowball Earth'.

**Another speculative astrophysical aside** (of Satan falling like lightning):

I refer again to Tom Findlay's very provocative book *A Beginner's View of our Electric Universe*. The idea that all the meteorite craters we see may be from arc plasma events is based on his book, and his description of their potential and power is very thought-provoking. My interpretation of the fall of Satan and the angels by 'lightning' is based on his observations of electrical discharge machining (EDM) as used in industry today.

I visualise a large, irregular 'solid block' of arc plasma lightning gouging out the Earth's surface and picking up billions of tons of debris (rather like a tornado) to produce the sometimes vast (two miles thick) diamictite layer that appears in the Snowball Earth rock strata. I can also visualise the electromagnetic discharge being powerful enough to transport boulders thousands of miles away (or even into space), and to drop them back to Earth, producing the exploding boulder and rivers of stone effect seen in the rock strata at this point in the stratigraphic column.

I would expect God to 'aim' not to kill any 'nephesh chayyah' animals (any subsequent follow-on catastrophes likewise), and would expect the strata thus produced to not feature intact flora and fauna but maybe some remnants of non-'nephesh chayyah' animals and plants pulverised and fried by the process.

The potential differences between this lightning and other charged bodies (planets) as they pass by may lead to discharges between them, perhaps explaining some of the concentrated cratering on these planets. This also may give an answer to mystery of the asymmetrical distribution of craters on these planets.

**Speculative astrophysics of Satan's first reaction:** I believe that Satan had some sort of governance role on Earth, in the physical realm (and likewise, the angels had governance roles too), and also throughout the Cosmos, with a specific dealing with gravity. 'Satan sealing up the sum' in Ezekiel 28:12 is a reference to this knowledge and understanding. [RDB] Angels also have specific stars allocated to them; hence they are sometimes referred to as 'stars' themselves. It could be that a relaxing of the precise gravitational parameters around planet Earth enhanced natural catastrophes, playing a role in the Snowball Earth and the subsequent flooding, leading to the Cambrian layer. It is also conceivable that Satan had more power before he was cast down. Satan immediately tries to destroy Earth as soon as he discovers the plans for Man, and this is before his fall. Psalm 8 and Hebrews 2 are describing Satan venting his anger and jealousy in God's courts. 'What is man that thou is mindful of him?' is spoken by Satan himself. Now, back to the history of everything...

**Non-single-cell fossilisation begins**

The perfect ecology of God's creation is disturbed, and for some flora and fauna, extinction beckons, as they may now lose their habitat forever.

This event is the first catastrophe and it is worldwide: the Snowball Earth disaster appears in the stratigraphic column, and so it corresponds to the fall of Satan and the angels.

Much slow flooding occurs from this catastrophe, with excess water from the perforated water shield and the resulting general slow rise in sea levels meaning there is no longer a calm sea. Huge amounts of potential sediment built up over millions of years are quickly leached away from and through the floating land mats, and this buries very many small organisms above, around and below. The coastline moves inland.

The Cambrian fossil record is laid down and dominated by the organisms unable to escape this slow ecological catastrophe situated in and around these floating mats. As a result, shallow marine flora and fauna and mud dwellers in their trillions dominate the fossil record at this time. God's staggering diversity and abundance of creation in this habitat becomes apparent. The subsequent effects are confined to non-'*nephesh chayyah*' organisms (animals without a spirit or soul) by location and ecosystem.

## The Palaeozoic Era

### The Earth groans, wrecking-ball Satan and the Permian extinction

Eventually, Satan and his angels work their way to the Earth's surface with a will to destroy Man's home-to-be and all of life on Earth, so they begin to wreak havoc. Satan starts to impart evil into the world. He sets about making the whole world a wilderness by every means available to him (Isaiah 14:16-17[138]). It may be that earthquakes, volcanism and the resultant tsunamis are his primary weapons – as we know he made the '***earth to tremble***' (Isaiah 14:16). These weapons may be multiple earthquakes as he tries to wipe out the whole world.

He takes control of the weather (the Prince of the Air), so spiritually induced catastrophes begin (Ephesians 2:2). (Satan may even be able to facilitate or orchestrate meteorite or EDM events, or the destruction of other planets. There is some evidence that planet destruction has

[138] Isaiah 14:16 (KJV): ***They that see thee shall narrowly look upon thee, and consider thee, saying, Is this the man that made the earth to tremble, that did shake kingdoms***;

occurred in the solar system.) Vast swathes of flora and fauna are decimated, and the presence of evil changes everything. Sin enters the world and the whole of creation groans (Romans 8:22).

A double onslaught is hitting planet Earth – not just that from Satan, but also a reaction to sin from within the creation itself (the creation '*groans*') with further extreme weather, EDM, extensive volcanism and meteorite activity. The delicate balance and perfect harmony of the creation suffers ecological catastrophes as rain falls for the first time, with erosion, flooding and sea surges leading to sea-level rises. The coastline moves further inland. The dense mist of the atmosphere thins.

The perfect stability is lost. Watcher Angels protect 'higher' '*nephesh chayyah*' animals; the catastrophes are not generally in their locality and not severe or quick enough to stop an escape. They therefore do not appear in this earlier fossilisation process.

The multiple catastrophes and recoveries producing different layers within the strata account for the multiple periods within the major eras. Successive ecosystems are destroyed. Organisms subsequently diversify within their inbuilt genetic potential; epigenetic switches are flicked by the rapidly changing prevailing conditions.

**The footprints in the sand**

As the flooding process is sometimes slow, at first, some animals can escape, so their footprints are found initially in the fossil record before they themselves succumb.

Due to the rising sea levels and a now rough sea, the catastrophes start from the coastline and work inwards. Plants cannot escape this process, but many animals can, so we see a gradual 'terrestrialisation' of plants in the fossil record, starting with marine plants and marine-side plants, and then progressing to those found further inland. Maritime and shoreside habitats are flooded first. Further vast amounts of sediment are washed away, burying organisms in that vicinity. More animals and plants located in these regions die, and, due to a permanent loss of habitat, some are never to be seen again.

The mass fossilisation in the stratigraphic column is largely by habitat and location, due to the larger animals and birds still escaping by their superior mobility. Floating mats of vegetation are decimated and no longer widespread, and their delicate trees are lost to the water (in their trillions), which leads to vast coal deposits being laid down.

Some further catastrophes (whether they be by direct action from the fallen ones, the Earth groaning or follow-on effects of previous catastrophes) result in death and fossilisation of flora and fauna from the affected ecosystems. The dominant animal and plant types will dominate the fossil record, and the changing ecosystems will show different dominant types; this too will show in the strata.

### Blurred lines

With more fossils becoming available, I would expect further blurring of these boundaries as the rarer organisms that have been there all along start to be found. So, some organisms will be rendered extinct, while others will change within their own kind.

Roger Sanders and Steven A. Austin give a very good explanation of coal formation in their floating mat model.[139] I suggest it could be that this process was augmented by multiple floods.

### Carnivores

I believe Satan had to have 'consent' to possess the snake[KK] so, in consistency, Satan could not initially control the higher '*nephesh*

---

[139] R. W. Sanders and S. A. Austin, 'Palaeobotany Supports the Floating Mat Model for the Origin of Carboniferous Coal Beds', Proceedings of the Eighth International Conference on Creationism, ed. J. H. Whitmore, pp. 525–552, Pittsburgh, Pennsylvania: Creation Science Fellowship, 2018.

*chayyah'* animals. The wages of sin is death, and as sin spreads through the animal kingdom, death, disease and decay also spread as the whole of creation 'groans' more. Carnivorous activity by '*nephesh chayyah*' animals in the fossil record corresponds to sin entering this 'level' of animals at that point, and shows how far up the hierarchy of these animals Satan has penetrated. Carnivorous activity and predation may also be a reaction from within the creation to the presence of sin – this is all built into the initial genetic design – and is also an example of the creation groaning. It may be that predation symbolises a corruption of Man's spiritual gifts, and this may also be a reversible dynamic as we know '***the wolf shall dwell with the lamb***' (Isaiah 11:6).

### The Permian extinction

These processes continue for a long time, but eventually sin becomes pervasive throughout the world and the Watcher Angel protection may be removed from more animals (and possibly 'higher' animals). One such particularly severe meteorite event (or electrical discharge mining event), with the resultant volcanic activity, results in the Permian extinction. This leads to mass extinctions with as much as 95% of sea animals and 70% of terrestrial species being wiped out too. Some animals protected by Watcher Angels can escape this, as can others protected by location and mobility, and those capable of adapting to the new prevailing conditions.

The world continues to re-establish itself after disasters for many years, with different dominant animals showing changes from inbuilt genetic potential; this process of destruction and re-growth continues for a long time. The fossil record supports this. (It may be that the Permian extinction coincides with Satan eventually reaching Earth's surface, and the preceding fossilisation is all due to the Earth groaning; this is not my position, though.)

## The Mesozoic Era

**God's first judgement is coming: darkness and a flood – the Cretaceous-Tertiary boundary** (also known as the K-T, Kt or Cretaceous Paleogene (K-Pg))

This disaster-reestablishment process continues but God's patience eventually runs out, and He judges the world by removing His Glory and plunging the Earth and also (by 'switching off the electric current') the whole Cosmos into darkness. All life on the planet perishes with a catastrophic and violent flood (with the possible exception of non-photosynthetic microbes). He reaches from Heaven and bursts the water vapour shield with His finger and the Mesozoic era ends.[KK]

Trillions of gallons of water fall at great speed onto the Earth, splitting land masses, inverting them and catapulting them over one another, with some land masses finishing many miles from their original position, burying much flora and fauna almost instantly in the process. Protection from Watcher Angels is removed and now all animals and plants are wiped out, with many new organisms appearing in the fossil record. In some areas, significant amounts of the atmosphere are also trapped by this process (Isaiah 24:1[140]). The fossils of chronologically coincident organisms appear to be from different eras in the fossil record as a result of this inversion and burying (Psalm 104:6-7[141] and Job 9:5-6[142]).

---

[140] Isaiah 24:1 (KJV): ***Behold, the LORD maketh the earth empty, and maketh it waste, and turneth it upside down, and scattereth abroad the inhabitants thereof.***

[141] Psalm 104:6-7 (KJV): ***Thou coveredst it with the deep as with a garment: the waters stood above the mountains.***

[142] Job 9:5-6 (KJV): ***Which removeth the mountains, and they know not: which overturneth them in his anger. Which shaketh the earth out of her place, and the pillars thereof tremble***

This is a different flood from that of Noah; it is much more violent and quick, and comes from above only. The power from God's finger may even leave a crater imprint on Earth.

Massive burying of animals and plants occurs, with huge amounts of mud deposited extremely quickly and in layers. Again, millions of years of potential sediment, this time, further inland, than has never been affected by the previous catastrophes is suddenly swept away. Much more than Noah's flood and unlike Noah's flood, there is no scalding water component from the depths to disintegrate some of the animal and plant matter. Organisms are fossilised in order by habitat, region and their ability to escape the flood (their mobility). There are no humans, so no human fossils are present. Several studies show a flood will result in a deposition of layers, acting a bit like a sifting system – Dr Andrew A. Snelling has discussed this.[143]

The Mesozoic Era fossil record is thus laid down and completed.

This is the first world judgement; it has a significant parallel with the fall of Adam, and it will also mark the boundary for a change to all animal and plant life that reflect life under the Abrahamic Covenant – how the world is watered, how the world is lit will both reflect where Jesus Christ 'is'.

Several studies have supported the Cretaceous-Tertiary as a worldwide flood boundary.[144]

---

[143] A. A. Snelling, 'Layers of Sedimentary Rock Deposited by Global Flood', Answers in Genesis, 2020, available at https://answersingenesis.org/the-flood/flood-cataclysm-deposit-uniform-rock-layers/ (accessed 10 May 2023).

[144] M. J. Oard, 'Is the K/T the Post-Flood Boundary? Part 1: Introduction and the Scale of Sedimentary Rocks', *Journal of Creation,* vol. 24 (2), 2010, pp. 95–104, available at https://creation.com/kt-boundary-flood-1 (accessed 28 November 2025).

J. P. Tomkins and T. Clarey, 'Palaeontology Confirms a Late Cenozoic N-Q Flood Boundary', Institute of Creation Research, 2020, available at https://www.icr.org/article/paleontology-confirms-late-cenozoic-flood-boundary (accessed 10 May 2023).

### A dark ice ball epoch

God continues His flood judgement by breathing on the water and freezing the now totally water-covered Earth (Job 37:10[145]). The switching-down of the electric universe leads to darkness throughout the whole Cosmos with no new light. In addition to the light switch-off, no light can propagate at all on planet Earth due to *'a thick swaddling band around it'* as described in Job 38:9. These judgements last for millions of years (Isaiah 45:12, Jeremiah 4:23 and 28[146]).

A vast amount of rock formation occurs as a result of the enormous pressures involved, giving an illusion of aeons of time, but in reality much of the rock formation happens relatively quickly. The Earth (surface of the water) remains frozen throughout this entire period. No animal or plant life survives – a dark ice ball epoch. The deep freeze preserves some organic tissue for a very long time. The Earth remains in a state of constant seismic activity for a considerable time after this judgement (Psalm.82:5-7[147]).

**This event marks a very significant boundary and is presented as the KT extinction level in the stratigraphic column. This will correspond to Genesis 1:2** – a flooded and dead world in darkness and with the waters frozen to some depth (*'the earth was without form and void and darkness was on the face of the deep'*).

---

[145] Job 37:10 (KJV): ***By the breath of God frost is given: and the breadth of the waters is straitened.***

[146] Jeremiah 4:23 (KJV): ***I beheld the earth, and, lo, it was without form, and void; and the heavens, and they had no light...*** [28] ***For this shall the earth mourn, and the heavens above be black: because I have spoken it, I have purposed it, and will not repent, neither will I turn back from it.***

[147] Psalm 82:5-7 (KJV): ***They know not, neither will they understand; they walk on in darkness: all the foundations of the earth are out of course. But ye shall die like men*** *(so they may not be men),* ***and fall like one of the princes.*** *(Fallen angels on a judged earth.)*

**Another speculative astrophysical aside of a dark ice epoch:** The whole Cosmos is in darkness, and the Earth also seems to have a deeper and thicker darkness. I suggest God turns down the 'capacitor', so the entire electric universe current is switched off, all plasma goes to dark mode and the stars give off no new light. The ZPE strength is greatly reduced throughout the entire Cosmos, except for planet Earth. If Jesus 'stretching out the heavens' has a meaning of continually stretching, then I believe this, too, stopped at this point. Now, 'darkness' is everywhere:

Job 38:9 (KJV): ***'When I made the cloud the garment thereof, and thick darkness a swaddling band for it...'***
Exodus 10:21 (KJV): ... ***'that there may be darkness over the land of Egypt, even darkness which may be felt'***.

Earth has an additional judgement as well, so that no light can shine on Earth. God makes a thick band of highly disordered ZPE, preventing the propagation of any light whatsoever. Atoms under this influence achieve maximum mass due to the enormous jiggling forces by the now extremely irregular ZPE, and this weight can be 'felt' – 'a darkness that you can feel' (perhaps due to the heavier atomic masses?).

Atomic clocks tick at their slowest possible rate, and atomic decay is very slow, much slower than at present. Atomic energy is at its lowest. If light could propagate, it would be extremely red-shifted. Satan and his angels are trapped in darkness – they cannot make light themselves, as light itself cannot be produced by any process; the light is 'trapped' in the atom.

But God wants a family and decides to make Man. He wants him to be made higher than the angels but from a 'deeper' (the '*lower*' in Hebrews 2:7) part of God Himself. He made the spirit aspect of Man by a curious process He likened to 'wroughting' ('made or fashioned in a specific way' or 'shaped by beating with a hammer').

## The Cenozoic Era

**The reinstallation and** *'The beginning of the creation'* (Genesis 1:3-31)

I believe angels are made in the image of God's holiness, and Man is to be made in the image of God's love[RDB] (Psalm 139:15[148]).

He decides it is time to selectively remake many parts of the original creation and to populate it with a very special new creation – Man – made in His image and His likeness. He carries out a 'reinstallation programme' with organisms copied from the previous creation: some copies, some modified, and some changed so much as to be called new creations. The previous creation *'brings forth'* organisms with the same or a similar design that used to be present in that first world (*'bring forth'* is a design acknowledgement phrase, a kind of 'copy and paste' or edit of a 'memory' from the previous world). The design is somehow carried in whatever medium is doing the 'bringing forth'. God almost seems to issue a copyright acknowledgement as He 'calls back' designs from the first creation, but does not call back all of the previous world's flora and fauna. The distribution, size and type of animals and plants will satisfy a metaphor for born-again Man's spiritual gifts.

He does this in six literal 24-hour days if viewed from planet Earth.

God instantly melts the ice on the still-frozen Earth, and God's Holy Spirit moves upon the *'waters'*.

**Day 1: A literal 24-hour day as viewed from planet Earth**

God lets light from the spiritual realm be visible in the physical; the light is similar but a lesser imitation of His Glory Light. This light fulfils a 'type' – a parallel meaning comparing Man's spirit with planet Earth.

---

148 Psalm 139:15 (KJV): ***My substance** (Man's spirit) **was not hid from thee, when I was made in secret, and curiously wrought in the lowest parts of the earth** (made from God Himself).* When did this event take place? Is it at conception or before then?

This may give one reason as to why light itself possesses a mysterious duality within it, as this fulfils the 'type'; as previously mentioned, God the Father (light), the Holy Spirit (waves) and the Lord Jesus Christ (photons). Unlike God's Glory Light, it is now under a command to be separated from darkness, so it is not as pervasive and there can now be darkness and light too.

The light becomes instantly visible in the entire Cosmos; its origin is in the spiritual realm. I believe Moses is taken back in time to view the whole process in order to write Genesis. Earth is a lesser version of heavenly things, including light. I speculate that the spiritual realm light 'loses' something in the process.

**Day 2: A literal 24-hour day as viewed from planet Earth**

God separates the water into two parts: water on Earth and water held in the sky between two glassy-looking, very strong membranes as in the first creation (Job 37:18[149]).

**Day 3: A literal 24-hour day as viewed from planet Earth**

God rearranges the land and water so that land is visible as one land mass. The omission of the words '*bara*' and '*asah*' in Day 3 reveals that God makes vegetation based on and similar to – but not necessarily identical to – vegetation in the previous world. No newly created designs of vegetation were used at this point, and some of the previous world's vegetation is not copied and remains extinct. For every plant in our present world, the design is present in a plants in the previous world. Any plant not yet found in the fossil record will be, and some flowering plants will look more like trees in the fossil record, being 20 feet tall or higher. The remade trees are now smaller (in parallel to saved Man's spiritually gifted knowledge being smaller than Adam's).

[149] Job 37:18 (KJV): ***Hast thou with him spread out the sky, which is strong, and as a molten looking glass.***

Some plants may originate from Eden and be new designs. There is biological similarity with the plants of the Mesozoic Era (Job 26:5[150]).

**Day 4: A literal 24-hour day as viewed from planet Earth**

The Word 're-fires' up the Sun, and moves the Moon and/or Sun and Moon to their present locations and into their current orbits. God re-ignites the stars by turning up the 'dial' and restores the charges to the electric universe in the entire Cosmos. Light from some, but not all, of the stars almost instantly reaches Earth (*'and it was so'*). The Earth is now dually lit (Isaiah 45:12 and Isaiah 48:13[151]).

**Day 5: A literal 24-hour day as viewed from planet Earth**

The use of the words *'bara'* and *'which the waters brought forth'* in this day shows that God makes all aquatic life, birds and flying insects based on designs from before the KT extinction. Some are close copies, some have modifications and some are so different as to be termed new creations. The design for these creatures is somehow held within the waters and 'called back' by God. This is not limited to sea-dwelling creatures, but to any creature whose 'design' is 'held' there. God makes a new creation of *'whales'* and many new birds at this point too. There is pronounced biological similarity to the fish, birds and flying insects of the Mesozoic Era, but also very significantly different new creations, albeit still based on previously existing organisms. This modern-created whale is both significantly similar to a 'whale-like' creature found in the fossil record before the KT extinction level, and, contrarily, significantly different (Job 26:5).

---

[150] Job 26:5 (KJV): ***Dead things are formed from under the waters, and the inhabitants thereof*** *(God is using the design of previously created living organisms from the first world).*

[151] Isaiah 48:13 (KJV): ***Mine hand also hath laid the foundation of the earth, and my right hand hath spanned the heavens: when I call unto them, they stand up together*** *(stars light up).*

### Day 6: A literal 24-hour day as viewed from planet Earth

The use of the word *'asah'* and *'it was so'* here shows that God may remake animals and insects based on a previously existing design from the original creation before the KT extinction event, but this could include new 'creations' based on these too (*'asah'* is anything God has worked on and *'bara'* a whole new design – making *'bara'* a subset of *'asah'*). There is pronounced biological similarity with animals of the previous eras (*'beast'* is an all-inclusive term for mammals and reptiles).

The text does not necessitate that every creature is brought forth; there may be many that remain extinct.

God makes Man and copies His own design (body, spirit and soul corresponding to Jesus, God the Father and the Holy Spirit) and the same morphology: head, body, face etc. He also makes him in His likeness, thereby Man having some of God's 'identity' and attributes. This design is an exalted high privilege, much coveted by Satan.

I believe God took four hours on Earth to make Man's soul.[KK]

### The same principles of design as the previous creation

The second world is of awesome beauty and in perfect harmony. Before Adam's fall, Man could see the spiritual realm with the physical, and hence God. Modern Man can only see the physical realm and just hear God.[KK]

There is no death of *'nephesh chayyah'* animals, no volcanism, no extreme weather and no meteorite (or EDM) impact. *'Nephesh chayyah'* animals grow and mature but do not age and degenerate. The reproduction rate is bounded. All present-day organisms and all organisms present in the Tertiary and Quaternary periods (or their predecessor kind) are present at this point. An abrupt change in the fossil record would be expected between the Cretaceous and the

Cenozoic, many new creatures suddenly appearing with biological similarity to the previous era's animals and plants, but with many significant differences. The fossil record confirms this is so.

The fossil record will show what proportion of animals and reptiles God made or chose not to remake. We can see that at this time of Adam and Eve, there is an almost complete absence of dinosaurs and reptiles compared with the previous Cretaceous epochs. Most dinosaurs were not remade, but many mammals, and some with significant differences, were added. This is consistent with '*beast of the earth*' and '*fowl of the air*' being made according to Jesus as a man's '*kind*' (a mammal). There is a gap of millions of years between the Cretaceous and the Cenozoic, but the slowdown of atomic decay caused by the increase in ZPE disorder may mask this gap.

The flora and fauna of the Palaeocene and subsequent periods may change, with their inbuilt genetic potential to do so. By similar processes as in the previous creation, some organisms may later become extinct and some will, by their genetic richness, diversify into the modern-day '*kind[s]*' we see now. There are new species in this creation, and true primates now make an appearance, but they will have to be based on an animal whose design is 'called back'. The world is now watered from the ground upwards.

**A speculative astrophysical aside of Genesis 1:3-31:** When God said in Day 1, 'Let there be light', light was instantly visible on and around Earth. I suggest that this is light from the spiritual realm. The 'swaddling band' around planet Earth is immediately changed from chaotic to ordered. In Day 4, I suggest, God switches back on the electromagnetic 'capacitor' for the whole plasma universe, and reorders and rebuilds the ZPE. The current for the electric universe being switched back on cascades throughout the Cosmos: an Earth-centric propagation (the Sun is lit before the stars – Genesis 1:16).

The stars start to shine immediately, but the ZPE and electric currents that feed the stars are of reduced strength, and they take time to build and spread throughout the Cosmos.

All the initial light emitted around the Cosmos is red-shifted to the maximum amount. The proceeding shift to blue is quick.

The speed of light is very quick and at its maximum, rather like the original creation, as the growing ZPE is now more ordered and no longer impedes the flow of light.

When we look at the very distant stars now, we are seeing starlight from before the ZPE was at its full extent, and some of these stars show this red-light shift process. The light from these stars is only just reaching here, and some light is still yet to reach here. The ZPE starts to increase across the Cosmos, but again geocentrically.

The speed of light is initially quicker than at present, as the ZPE is ordered, but when the effects of sin again hit the Cosmos, this leads to the ZPE 'groaning' and becoming more random and less ordered, hence greatly increasing its turbulence on subatomic particles and, in turn, slowing down the speed of light to today's pace.

**Sin wreaks havoc in the creation again: more catastrophes –** Adam falls, sin enters this world and the Earth groans. The light from the spiritual realm starts to fade. The same eternal rules apply; the Earth is now subject to extreme weather, volcanism, meteoric impacts and earthquakes (or EDM cratering). As a result, there are many disasters and much fossilisation occurs. Satan imparts evil into the creation again, but it is centred on Man.

Two main disasters are recorded in the Bible: Noah's flood, and the days of Pelag when the land mass was divided. I believe the land literally split and water flooded into the gap from below.[KK] Both disasters lead to many years of follow-on disasters (plus an ice age) and subsequent recoveries, all of which appear as mass fossilisation in the strata. There are several studies that suggest Noah's flood boundary is in the late Cenozoic.[152]

**A speculative astrophysical aside of Noah's flood:** In discussing Day 1, I described how God let light from the spiritual realm flood in; this is paralleling God lighting Man's path to salvation. But it is not clear when this light is completely switched off. It is not visible today. I am going to speculate that it started to dim at the fall and was completely switched off at the time of Noah's flood. Genesis 8:22 states: '***While the earth remaineth, seedtime and harvest, and cold and heat, and summer and winter, and day and night shall not cease***.' I speculate, again, that the conditions on Earth were much more stable before the flood, and that both the water vapour shield and light from the spiritual realm helped to achieve this. This passage defines the new, more volatile Earth, with more extremes of darkness and light and seasons. The '*seedtime and harvest*' means this is how Man will now live and be judged. I also suggest that these conditions may trigger many epigenetic changes, and all may well be representative of the salvation process. Within the KRM I suggest that '***seedtime and harvest, and cold and heat, and summer and winter, and day and night***' all have metaphoric meaning of before- and after-salvation or good and evil.

The presence of this 'spiritual' light (and the subsequent removal of it) may also influence atomic energies and properties in a similar manner to the previous world (but not to the same degree, as this is now shining on the Earth and not from 'within' it). This, therefore, could also explain some of the less severe dating anomalies in this era.

Another timeline for the whole world becomes apparent and explains where Jesus is: how this links with the origin of spiritual gifts

---

[152] J. P. Tomkins and T. Clarey, 'Palaeontology Confirms a Late Cenozoic N-Q Flood Boundary', Institute of Creation Research, 2020, available at https://www.icr.org/article/paleontology-confirms-late-cenozoic-flood-boundary (accessed 10 May 2023).

to Man; and how this in turn links the size and characteristics of flora and fauna on Earth to these spiritual gifts.

The first world had a water vapour shield, and two thirds of the water on Earth was in it (representing Jesus and the Holy Spirit). Mist cascades downwards from the canopy (representing Man's spiritual gifts as coming from these two, located in Heaven above but 'walking with Man' – Genesis 3:8). Jesus is God, and so all flora and fauna are large and dominated by reptiles. These are the conditions associated with unfallen Adam. Then a worldwide flood occurs, and then comes Genesis 1: I speculate that the water vapour shield then had one third of the water in it (representing just Jesus). The mist rose from the ground, which represents Jesus 'agreeing' to become a man while still located in Heaven. Jesus is God, so flora and fauna are still large but are now dominated by mammals (megafauna). Man's spiritual gifts are dependent on the promise of Jesus Christ becoming a man at this time.

These are the conditions representing life under the Abrahamic Covenant. Then, a worldwide flood occurs (Genesis 6: Noah's flood). There is no firmament; Jesus has now become a Man and sacrificed Himself for all sin; all water is still from the ground and all Man's spiritual gifts are from Jesus Christ because of His sacrifice as a man. All flora and fauna are now smaller and have different characteristics to reflect Jesus as a man. These are the conditions under the better covenant with Christ. The first flood represents Adam's fall, spiritual death and judgement, and the second represents Christ's death and resurrection. Perhaps a water vapour shield (Jesus) holds the mist (the Holy Spirit) on Earth?

Please note that the entirety of this timeline may seem evolution-mimicking in appearance. I see no reason why the animal size and traits may not sometimes be a dynamic with incremental changes rather than static. To repeat, I believe that the Genesis days are literal 24-hour days, but God may bring animals that existed before the Cenozoic to a certain 'state' to then 'bring them forth' in order to remake them in Genesis 1 (likewise animals before Noah's flood). Some of these changes within animals may even 'need' to be over time to correspond to a time under the Abrahamic Covenant, or even to correspond to Jesus' life's work in achieving what these changes metaphorise. (A little bit of theistic evolution! Certainly, enough to look like evolution.) I do not see God surgically intervening on existing genomes, but rather epigenetic switches flicking and inbuilt design being realised. I speculate that the triggers for these may be atmospheric pressure, the presence or absence of God's Glory Light, thick moisture (present or missing in the atmosphere), time itself, generational counts, zero-point energy degradation or sin. (The last two may even be interrelated.)

I suggest that species diversification occurs as soon as creation happens in the Genesis six-day account (representing spiritual life under the Abrahamic Covenant), and I also suggest that much more rapid and wide-ranging species diversification (to represent the better covenant of Christ) occurs after Noah's flood. In effect, the uses of the word 'replenish' in Genesis 1:28 and again in Genesis 9:1 are references to the same event in their literal meanings – both restocking an empty earth. In the KRM God is telling Man to restock his 'empty' spirit. The ever-burgeoning and diversifying new life testifies to Christ's better covenant doing the same in Man's spirit. Nor do I see a reason why God did not make horses, elephants, livestock, cats and many more animals change within their 'kind' before Noah's flood, and then the subsequent changed animals diversify even more after that.

It may be that the Cambrian widespread death of 'creeping things' and 'non-blood living creatures' signifies Adam's initial loss of interaction with the Holy Spirit (thus, the atmospheric mist may do the same and 'thin down' in its nature).

But there may be even more possible parallels. Interestingly, the first world is created, then has a first major catastrophe (Snowball Earth) and a second major catastrophe (the Permian), before dying at the KT extinction level. Our present world too has had two major catastrophes: Noah's flood and the days of Pelag. Man is born and then dies, but it appears that he too has two attacks on his body when ageing occurs more than at any other time in his life: one around age 44, and another around age 60.[153] As the last two are instigated by God, this raises the possibility that the Permian may be a judgement from God, although no obvious consistent time ratios are present.

**The population scatters**

In the time of Babel (about 4,000 years ago) new ethnicities and languages are introduced.[KK] God is proactive in this process, and it may be that human epigenetic switches after the flood play a role in this too. (It may even be that humans' physical appearance becomes more varied after Noah's flood, rather like animals diversifying; this is consistent with more ethnic variety.)

The widespread catastrophes result in widespread loss of human structure and society, so, while rebuilding, many societies must live more primitively than before. Caves and temporary shelters are used, and archaeologically these give the appearance of a less advanced culture than there actually is; these people groups are often termed 'cavemen'.

**God's contingency plan – diversification – change is a constant**

I believe that God built within His genetic plan a contingency for any eventuality; sin and death entering the world included. The genetic

---

[153] This research was pointed out to me by Suzanne Parsonage in a conversation in August 2024:

X. Shen, C. Wang, X. Zhou et al., 'Nonlinear dynamics of multi-omics profiles during human aging', *Nature Aging*, 4(11), pp. 1619–1634, 2024, available at https://doi.org/10.1038/s43587-024-00692-2 (accessed 28 November 2025).

change towards predation involving '*nephesh chayyah*' life will begin at this point, but this design potential was already in every organism.

**Disease and quick modifications**

The 'fallen state' that all creatures are born into after the fall impinges on all genomes. Bacteria, viruses and other pathogens must have changed in character or location to become pathogenic (or the immune systems of animals weaken and let these microorganisms become pathogens, or both). Again, the whole creation groans. This makes disease also an inbuilt potential, realised with the corruption of genomes and weakening of the flesh – sin-triggered epigenetics?

**Quick diversification** (being '*fruitful*' – and huge genetic potential)

God's creation constantly changes, and creatures constantly change, with many inbuilt variations coming to the fore but always bounded within their own '*kind*'. Some kinds may lead to many thousands of species. Genetically rich core animals in the arc, with their inbuilt diversification potential, diversify into all modern animals.

**'God still does it'.** God's blessing to '*be fruitful and multiply*' is, I believe, an ongoing blessing: not a 'one-off' but more a statement of continuing care, and a genetic gift for animals to be able to diversify, and this at the same time will symbolise Man's spiritual condition. The mechanism of this blessing will likely be unfathomable and seamless but will, I suggest, bring with it the selection of the best genotype and phenotype prodigy to be successful in the prevailing conditions and ecosystems. A little nudge or twist here and there; in other words, God 'favouring' something.

I believe God will have selected the most genetically appropriate animals for the Ark: always the right genome, right place, right time – a Lord of logistics or perhaps a Lord of 'nano'-logistics, or maybe even a digital Lord of Planck length logistics and zeptoseconds.

I picture DNA in a nucleus looking like a barrel full of very long earthworms wriggling around all over the place, and some seemingly random bits breaking off and rejoining somewhere else. But for all mutation scenarios, they are not really random at all – we just do not recognise the order yet. I believe multiple genetic changes may occur, way beyond chance mutations, all to produce a new species within the same kind. And yes, God remains the project manager.

Could random mutations produce these balanced ecosystems changing together, or new proteins that happen to match up with other proteins? Is beauty just a lucky side effect? The whole creation is 'uttering speech' with its changeability and constant beauty. What a work this creation is! So, yes, God intervenes, He did do it and He still does do it, but I do not know how, and I cannot see the join where He chooses to intervene. God is in those genetic details.

**Summary**

**How old is the Earth, then?** The Bible does not seem to say how old the Earth is, or perhaps we have just not found this yet. I believe the Earth's age and the gap duration is millions of years, but not tens of millions.[KK] So, not really a 'very old Earth' and not a 'young Earth', perhaps just 'a little bit of an old Earth'. Several million years? This would make the radiometric dating out by a factor of about one thousand (I am reminded of 2 Peter 3:8, '***… a day is like a thousand years…***').

**Evolution is debunked and now replaced.** So, by collating all creation theories proposed by many brilliant minds and never doubting the Bible's inerrancy – along with being open-minded to possible new hermeneutic interpretation and accepting all true science – and, finally, by paying close attention to the detail of revelations by Kerr and Bullock, I describe and explain…

A creation that thrives for millions of years, then experiences a catastrophe of Satan 'letting fly' at planet Earth, and a subsequent worldwide arc plasma event of Satan and the angels being cast down. This leads to a follow-on slow ecological disaster affecting mainly shallow marine and shoreside flora and fauna. Satan continually attempts to make the world a wilderness as he introduces evil into the world. He moves up the hierarchy of '*nephesh chayyah*' animals and uses all weaponry available to him; multiple catastrophes and subsequent recoveries hit the planet.

Eventually, God's patience snaps and there follows a whole world destruction by a flood-and-darkness judgement. The darkness comes with a universal shutdown and lasts for millions of years. This is marked as the KT extinction event. (I prefer **JC distinction event!**)

All organisms after the KT extinction are part of a reinstallation programme by God to house His most important creation – Man. Animals and plants are copied from the first creation: some the same, some with modifications, and some different enough as to be classed as new creations. The new 'remake' of genetically rich core animal and plant kinds also suffers many catastrophes and recoveries.

Although I concede that some of this is speculation, I still claim that 'the Long Life – Long Gap – Genesis Six-days Model' is the answer to all 'problems': all the strata, fossils, radiometric data and apparent great ageing are explained. Red-shifted light is covered, along with distant starlight reaching us – all are accounted for. Young-earthers, day age creationists and theistic evolutionists are all refuted by the Bible itself.

**The Long Life – Long Gap – Genesis Six Days Model**

**(The perfect spirit – fallen spirit – born-again spirit model)**

**Young Cenozoic (born-again) creationism**

**Or more succinctly: Born-again Creationism**

# The Final Chapter

## *'It's time'*

### The default is God

Did you know that God placed a knowledge of Himself in your heart (Ecclesiastes 3:11[154] and Romans 1:20[155])?

When you look at the stars at night, you instinctively know there is a God (Psalm 19:1-3[156]).

When you study the creation, you see and recognise and know it is His handiwork (Romans 1:20).

God makes Himself known to you many times in your lifetime.[KK] *'Here I am, do you want me?'*

You have repeatedly rejected these advances if you are not a Christian, God approaches everyone multiple times.

---

[154] Ecclesiastes 3:11 (KJV): ***He hath made every thing beautiful in his time: also he hath set the world in their heart, so that no man can find out the work that God maketh from the beginning to the end.***

[155] Romans 1:20 (KJV): ***For the invisible things of him from the creation of the world are clearly seen, being understood by the things that are made, even his eternal power and Godhead; so that they are without excuse.***

[156] Psalm 19:1-3 (KJV): ***The heavens declare the glory of God; and the firmament sheweth his handywork. There is no speech nor language, where their voice is not heard.***

Every child beneath an age of accountability is in Heaven. Every miscarried and aborted foetus is offered Christ in Heaven.[KK]

Many people are in Heaven who have never heard of the Bible. They can know God without knowing the Bible.

God does not send anyone to Hell – you send yourself there by rejecting Christ (John 12:48[157]).

To reject Christ, you must have been offered Him.

There is a Hell. *'If everyone went to Heaven then it would be just like Earth'*.[KK]

**Judgement**

Your every thought, word and action will be reviewed. This would not be a comfortable watch for any of us. Your whole life will be instantly known to all present in the briefest moment, and at this judgement they will all say, 'Yes, we understand and this is a fair and righteous judgement'. They will realise that for some of you, no matter what God did, you would not have bowed, even if He performed a miracle in front of your eyes (Luke 11:29-30[158]). You would bring sin into Heaven. God will not allow that.

*'At the time of your judgement it will not be a negotiation'* – Chuck Missler.

---

[157] John 12:48 (KJV): ***He that rejecteth me, and receiveth not my words, hath one that judgeth him: the word that I have spoken, the same shall judge him in the last day.***

[158] Luke 11:29-30 (KJV): ***And when the people were gathered thick together, he began to say, This is an evil generation: they seek a sign; and there shall no sign be given it, but the sign of Jonas the prophet. For as Jonas was a sign unto the Ninevites, so shall also the Son of man be to this generation.***

**God is righteous**

He cannot just write off your sins. He has to balance the books.

In some deep way, if God sent His only son to live a perfect life and die a voluntary innocent death, that can be used to write off your sins.

Jesus did not just die an innocent death, but took on a 'cup of trembling' – He took on board all sin (that could ever be committed by all men) and its due punishment, and that too was 'nailed' to the cross and crucified – that's how He can pay for your sins. He knew how hard it was going to be and asked the Father if there was any other way. (The anxiety beforehand was so much to bear that He sweated blood.)

All you have to do is accept this gift. He would do it all again, and He would do it all just for one of us – **He would do it all just for you.** True agape (John 3:16[159]).

**The Bible comes with a promise**

If you humbly seek the truth and you read the Bible, it will change you (Isaiah 55:11[160]).

**Life abundant**

It is the worry of the perceived 'boring' consequences of becoming a Christian that seems to stop some people from taking this step, but the fact is you will have life more abundant if you do (John 10:10[161]).

---

[159] John 3:16 (KJV): ***For God so loved the world, that he gave his only begotten Son, that whosoever believeth in him should not perish, but have everlasting life.***

[160] Isaiah 55:11 KJV): ***So shall my word be that goeth forth out of my mouth: it shall not return unto me void, but it shall accomplish that which I please, and it shall prosper* in the thing *whereto I sent it***

[161] John 10:10 (KJV): ***The thief cometh not, but for to steal, and to kill, and to destroy: I am come that they might have life, and that they might have it more abundantly.***

**An eternity**

I hope I can have a meal, play some sport and have a drink and a joke with you in eternity. No time, no pain, no suffering, all sports, the arts and just about all non-sinful activities are there and enhanced. I challenge you to read the book of John with an open mind and with a prayer to have the truth revealed to you. Ask God whether you need a saviour. God never lets you down.

*'The more you find out about God the bigger He becomes'*.[KK]

*'The more I know about the Bible the more literally I take it'* – Chuck Missler.

God did not make diseases, genetic abnormalities, carnivorous activity, extreme weather, volcanoes, earthquakes, asteroids, death and suffering. None of these were in the original creations; these are all an inevitable consequence and a reaction from within the creation to the presence of sin.

**The prodigal atheist**

Even though you have ignored God's many advances, even though you may have campaigned against Him, and even though you may have played a role in the eternal demise of others, God still waits patiently at the window, looking longingly up the road waiting for His prodigal atheist to come home. If you do, He will embrace you, bells will be rung in Heaven, and He will completely forget all of your previous sins; it will be as though they never happened. And then, when you use your gifts and skills, no matter what they are, big or small, when you use them to the full, then, you too, will *run fast and feel His pleasure*.

# Endnotes

---

[i] **Kat Kerr**: *Revealing Heaven I: An Eyewitness Account*; Jacksonville. Florida, USA: Xulon 2007, 2010. Kat Kerr, *Revealing Heaven II: An Eyewitness Account*; Jacksonville. Florida, USA: Xulon, 2010.
https://www.revealingheaven.com (Accessed July 2025).
https://www.katkerr.com (Accessed July 2025).

[ii] **Robin D Bullock:** https://www.robindbullock.com (Accessed July 2025).

[iii] **Flavin adenine dinucleotide (FAD) Man at or after Day 5**

A possible interpretation of the three six-vertices-containing-heterocyclic rings (see figure 6.1):

$CH_3$ may represent Christ's 'wood' and 'wine' as holding all together but also, the two $CH_3$ components may represent the two interactions with 'the three resources of Pison' – Man's spirit working in both 'Pison' and 'Gihon' with *'good gold'*, *'bdellium'* and 'onyx stone' ($CH_3$); and in other contexts, the $H_3$ of $CH_3$ may refer to 'Hiddekel', 'Assyria' and 'Euphrates'.

Ribitol can represent the working Holy Spirit in the same way as ribose.

The three rings from left to right are: a benzene ring (Man's image life given by God at salvation and what is now being worked on), a pyrazine ring (the likenesses acquired by Man) and a pyrimidine ring (the regenerated spirit).

Numbering from the top or apex vertex point first and going clockwise, and then to the next vertex, top first and clockwise again:
Benzene: (1) Beast of the earth; (2) and (3) the indwelling (in waters) of Jesus Christ and the Holy Spirit – both shared with the pyrazine ring; (4) the body of Christ; (5) the bride of Christ; (6) the image life from Jesus Christ that gives light.

Pyrazine: (2) and (3) the indwelling in waters of Jesus and the Holy Spirit – both shared with the benzene ring; (7) a likeness – N3; (8) the body of Christ; (9) the bride of Christ (8 and 9 both shared with the pyrimidine ring); (10) a likeness – N4.

Pyrimidine: (8) the body of Christ; (9) the bride of Christ (8 and 9 are both shared with the pyrazine ring); (11) the regenerated aspect of Man's spirit; (12) a likeness – N1; (13) beast of the earth; (14) a likeness – N2.

The 'N' allocation can change depending on context with four different sets of two 'N's. The two 'whole land' interactions, the two 'onyx stone' interactions, the two active likenesses of Pison, the two active likenesses of Gihon and the two likenesses of Hiddekel and Assyria (fruitful + multiply, replenish + subdue, fish of the sea + fowl of the air).

[iv] **Oxaloacetic acid ($C_4H_4O_5$) – oxaloacetate ($C_4H_2O_5^{2-}$) is achiral – Pison** – 'acid from the vinegar': Man's regenerated spirit's role carrying Christ-like attributes 'pleasant to the sight of God'. This is Christ centred and also centred on Man's regenerated spirit. To live up to its role in Day 5, this compound needs:

**C4** – All mentioned image life in Genesis 2:11: 'the four image lives of Pison'. The allocation of the carbon atoms can change in the conjugate base; the conjugate acid describes what is needed to yield the likenesses but the likenesses themselves contribute to: 'the body of Christ' (1); Man's original spirit continually being influenced (2); 'the beast of the earth' (3); and the regenerated Man (4). (Please note there is no 'bride' for Christ as that aspect is more a function of the Holy Spirit.)

**H4 and H2** – Regenerated Man gets credit for making four aspects of living waters flow as He is placed in the garden downstream of 'went out of Eden' and 'water the garden', so he just gets credit for 'parted into four heads – H4'. The regenerated spirit of Man has two individual outgoing actions that make Man 'pleasing to the sight' – 'Pison' and 'Havilah'.

**O5** – Oxaloacetic acid used four lights in Man: the lesser light, the greater light, the Bible and the Holy Spirit – but also continues to get more of Christ's light given (O5) – 'the five lights in Man'. The base oxaloacetate: the Holy Spirit works on Man's four lights to achieve his 'pleasant to the sight changes', and each of the two electrons hovers over two oxygen atoms in oxaloacetate in consistency – this includes 'the greater and the lesser light'.

[v] **Malic acid ($C_4H_6O_5$) – malate ($C_4H_4O_5^{2-}$) is L-chiral – Gihon** – 'the apple of God's eye' – Man (his whole original spirit now) in his role as 'the apple of God's eye' – carrying Christ's image life and being in 'the body of Christ'. This is Christ centred and also centred on Man's whole original spirit:

**C4** – Malic acid uses 'the four image lives of Pison' (the source of 'good gold', 'bdellium' and 'onyx', and Man's original spirit ('whole land')). The allocation of the carbon atoms can change in the conjugate base to: 'the body of Christ' (1); the original spirit of Man (2); 'the beast of the earth' (3); and the regenerated Man – being made (4). 'Gihon' is about changes in Man and please note again that there is no 'bride of 'Christ'.

**H6 and H4** – Man gets credit for making six aspects of living waters flow: 'went out of Eden', 'water the garden' and 'parted into four heads'. Man has four actions that make Man a working 'apple' – 'Pison', 'Havilah', 'Gihon' and 'Ethiopia'; malate has H4 in consistency with this.

**O5** – 'Man's five lights'. The conjugate base will reveal a $^{2-}$ charge associated with two oxygen atoms for each of the two electrons and this is consistent with the Holy Spirit working with four lights in Man including 'the greater light and lesser light'.

[vi] **Pyruvic acid ($C_3H_4O_3$) and pyruvate ($C_3H_3O_3^-$) is achiral** – the fire and the grape – the Bible with the Holy Spirit in roles as presenters of the Holy Spirt, Jesus Christ and Christ's fruit respectively. This is Holy Spirit centred and also centred on Man's original spirit. Its contributions in Day 5, in this day it needs or works on:

**C3** – For pyruvic acid, Genesis 2:19-20 tell us the Bible works with Man's whole spirit (or 'ground') – (1), Man's 'beast of the earth' (2); and the Holy Spirit's image life (3) – the source of the 'gifts of the spirit'. For pyruvate, the Bible's likeness contribution will continue to add to Man's original spirit, his 'beast of the earth' and now the *'bride' of Christ* (it is not contributing to the Holy Spirit's image life).

**H4 and H3** – The post-salvation Bible plays a role in the four river heads ('Mans four waters to flow') – H4', hence, pyruvic acid has H4. The Bible individually gets credited with three likenesses – 'Hiddekel', 'Assyria' and its part in 'Euphrates', hence pyruvate has H3 (Pyruvate has a specific timeline that includes a long-term aspect).

**O3** – The Bible still works in tandem with the two other lights: the Holy Spirit and the lesser light (there is no greater light included now as this is Holy Spirit centred). The Bible works together with the Holy Spirit hence pyruvate has one negative charge (the working Holy Spirit) over two oxygen atoms (the Bible and the Holy Spirit).

[vii] **Phosphoenolpyruvic acid is ($C_3H_5O_6P$) – phosphoenolpyruvate ($C_3H_2O_6\,P^{-3}$) – is achiral:** the light bringer, the purified substance, the fire and the grape. The Holy Spirit's role as a light bringer, a purifier of Man's spirit and a guide as to what is right and wrong. This is Holy Spirit centred and also centred on Man's original spirit. The Holy Spirit in Day 5 and limited to actions tied in with Hiddekel.

**C3** – For the conjugate acid the Holy Spirit will work with: Man's original spirit (1); the Holy Spirit's own image life (2); and Man's 'beast of the earth' (3). The conjugate base will represent the Holy Spirit being fruitful by changing: Man's original spirit (1); a 'bride' of Christ (2); and Man's 'beast of the earth' (3).

**H5 – H2**: The Holy Spirit contributes to five aspects of making living water flow (water the garden (1) and parted into four heads (2, 3, 4 and 5)) – 'God's five waters to flow'. The Holy Spirit (when not using 'kol' interactions and working under an 'Hiddekel' timeline) individually gets credited for two likenesses ('Hiddekel' and 'Assyria'); hence phosphoenolpyruvate is H2.

**PO4** – The Holy Spirit is a light deliverer bringing lights from a light bringer using four lights: the regenerated Man's spirit, Jesus Christ, the Holy Spirit and the Bible. The light bringer for these is Jesus Christ in order to form the waters, but the light bringer may be regenerated Man when evangelising with other men when using these waters. In the 'made to grow' verse of Genesis 2:9, it will be Christ as the light bringer; in the rivers of Eden it may reference regenerated Man as the light bringer, in which case the 'lights' in the 'light bringer' include the regenerated Man. In a different context it may be that God the Father is a light bringer too.

**O2** – The Holy Spirit uses two lights in Man in this fruitful work (the Bible and the Holy Spirit Himself). Each electron (the Holy Spirit) hovers between two oxygen atoms to represent working with all four 'light bringing' lights and the two lights in Man to explain the $^{-3}$ charge.

### Generic sweetness and light

There are at least 15 standalone-one-true-pattern 'sets of four' in Genesis 1 and 2. There are at least 11 sets of six in Genesis 1 and 2 and these hold three sets of four each (by the nesting process discussed in Chapter 13). I repeat the rationale here:

***'Be fruitful (1a),*** *and (sub category: multiply (1b)), and* ***replenish the earth (2a),*** *and (sub category: subdue it (2b)): and have dominion over the* ***fish of the sea (3),*** *and* ***over the fowl of the air (4),*** *and over every living thing that moveth (the long-term command for growth).*

*Because broad terms are used, some of these terms can apply to both the Holy Spirit and Jesus Christ; and similarities also exist in that '1a' and '1b' are very similar in nature to '2a' and '2b' (one going out of Man and one going into Man); also '2a; and '2b' are similar to '3' and '4' (both are additions and purifications); then by this understanding each set of six can be split into two further 'nested' sets of* one-true-pattern 'sets of four' *with categories '2a' and '2b' overlapping.* ***This overlapping also emphasises the shared works of Jesus Christ and the Holy Spirit*** *(in effect you can read each pattern simultaneously and then they are working together).*

*Fruitful ∞ multiply ∞ replenish ∞ subdue*

*Replenish ∞ subdue ∞ fish of the sea ∞ fowl of the air'*

This means that 2a and 2b can be 1a and 1b, and also, can be a 3 and 4. Genesis accomplishes this by using broad metaphoric terms for 2a and 2b (some of these are descriptions of spiritual conditions or considerations before salvation): *darkness + deep; signs + seasons; a great light + another great light; to rule over the day + rule over the night; replenish + subdue; no rain upon the ground + not a man to till the ground; breathed the breath of life + became a living soul; Gihon + Ethiopia; called every living creature + that was the name thereof: a rib + closed up the flesh thereof; flesh + my flesh.*

Many compounds in the biochemical KRM are based on sugar-as-a-metaphor for salvation (God in Man or the born-again Man), and many compounds metaphorise light bearing or light bringing (by containing phosphorous and oxygen). Compounds are often in the format (with '*ate*' meaning 'worked upon' or 'having the quality of'):
**phospho...gly....ate** or
**gly...phosph...ate.**

There are many different etymologically-lead metaphors to choose from with these compounds. Is the light going to or from the saved and 'sweet' spirit? Is the saved spirit acted upon or is the light acted upon? Are both the 'sweet' spirit and light acted upon, or are they both acted with? But please note in all six categories of a Christ's-role-expanded 'one true pattern', light goes in and out of saved-Man's spirit and light works on or with saved-Man's spirit too. Hence, these compounds show exactly the same versatility as the biblical terms and allow for the same multiple 'one true pattern' sets of four. Please note that there may be more than one interpretation that will arrive at the correct formula. For roles described in this book:

[viii] **3-phosphoglyceric acid ($C_3H_7O_7P$) – 3-phosphoglycerate ($C_3H_4O_7P^{3-}$) is D-chiral and is also known as 3-PGA, and 2-phosphoglyceric acid ($C_3H_7O_7P$) – 2-phosphoglycerate ($C_3H_4O_7P^{3-}$) is D-chiral and is also known as 2-PGA. 3-PGA is a '2a' in the Calvin cycle.**

**C3** – Conjugate acid: The Holy Spirit works with 'the three resources of Pison': Christ's image life (1); the Holy Spirit's image life (2); and Man's regenerated spirit (3). Conjugate base – this matures Man's: 'beast of the earth' (1); Man is more glued into the 'body of Christ' (2); and it all leads to more regeneration of 'Man's regenerated spirit'.

**H7** – 3-PGA has '$H_7$' in its acid form (Man's six waters to flow and just one of the likenesses has 'happened' in this river at this stage ('*Gihon*')) – '$H_7$'; and the base has '$H_4$' and this leads to the Holy Spirit working with the Christ-centred 'four active water likenesses' being 'sweet' 'Pison', 'Havilah', 'Gihon' and 'Ethiopia'.

**PO4** – The Holy Spirit is a light deliverer working with four lights: firstly in the acid, the regenerated Man, Jesus Christ, the Holy Spirit and the Bible (O4) give light to form Man's spirit; and secondly in the base, again the same four lights (O4) all work with Man to give light to the saved Man or other men – and so the light bringer for these may be Jesus Christ or regenerated Man respectively. The presence of 2 electrons hovering between the four oxygens is consistent with this.

**O3** – In addition to the four lights in a light bringer role-actioning salvation, the Holy Spirit also works on three lights in Man: the greater light, the lesser light and the Holy Spirit Himself. The presence of one electron is consistent with a representation of the Holy Spirit working with the light of Christ and it being given to other men.

The '3' can be a reference to working with entities: numbers seem to represent images, likenesses, waters or lights, but often images.

[ix] **Dihydroxyacetone phosphate ($C_3H_7O_6P$) – dihydroxyacetone phosphate ($C_3H_5O_3PO_4^{2-}$) – is also known as DHAP and is achiral:** a light bringer and a distilled product of vinegar; all giving waters and light to Man's spirit. The 'dihydroxy' appears to represent 'two water-bringing lights' (two 'O's) as a part of a light bringer (phosphate). This greater light ('O') of Christ is creating and ruling over goodness (day) and the lesser light ('O') ruling over the sin (night). In both cases it is the Holy Spirit doing the ruling as represented by G3P. In effect, G3P is the ruling itself.

**C3** – Conjugate acid: Christ's image life (1); Man's original spirit (2); and the regenerated aspect of Man's spirit (3). Conjugate base: 'the body of Christ' (1); 'the beast of the earth' (2); and the regenerated spirit (3).

**$PO_4$** – The Holy Spirit is a light deliverer bringing lights from a light bringer using four vehicles – Jesus Christ, the Holy Spirit, the Bible and God the Father – the light bringer for this is the Lord Jesus Christ. In the conjugate base two electrons hover between two oxygen atoms each, in consistency with this.

**H7** – The conjugate acid: this is representing Christ's life going into Man, this is Man's whole or original spirit and this contributes to all six aspects of making living waters flow (H6). The only likeness that is relevant is the 'conversion' of Man's original spirit to a regenerated Christ-like one – the whole ('kol') land of Ethiopia (H1): a total of H7. The conjugate base reveals what this process gives Man – the five likenesses of Gihon – H5 (with the three resources included) 'Gihon', 'Ethiopia', 'good gold', 'bdellium' and 'onyx stone').

**O2** – DHAP deals with two lights: the light of Christ (greater light) (1); the light of Christ in the regenerated Man (lesser light) (2).

[x] **Glyceraldyde 3 phosphate ($C_3H_7O_6P$) and ($C_3H_5O_6P^{2-}$) is also known as G3P and is D-chiral** – has two pathways, a '3' and a '4' in the Calvin cycle: For '3':

**C3** – Conjugate acid: The Holy Spirit's image life (1); Man's original spirit (2); and Man's 'beast of the earth' (3). Conjugate base: 'bride' for Christ (1); Man's original spirit (2); and 'beast of the earth' (3). (Genesis 2:19-20 describe 'fowl of the air' as being made of both the original and regenerated spirit. The 'bride' is a subset of the 'body of Christ'.)

**$PO_4$** – The Holy Spirit is a light deliverer bringing lights from a light bringer using four vehicles: Jesus Christ, the Holy Spirit, the Bible and the regenerated Man's – the light bringer for these is regenerated Man in 'the body of Christ' when Man is being 'fruitful' (the Holy Spirit using Man to evangelise – '4') or can be the Holy Spirit Himself when making Man spiritually grow – '3'.

**H7** – The conjugate acid: in '3', the Holy Spirit is represented by etymology and chirality in Hiddekel; this makes for contributing to five aspects of making living waters flow (5) and the two likenesses of 'Hiddekel' – 'Hiddekel' and 'Assyria' (H2) for a total of H7. The conjugate base delivers the likeness changes in Man: 'bdellium', His part as 'bdellium' in the 'cooperative', 'Hiddekel', 'Assyria' and 'Euphrates'.

For the second pathway of G3P in '4': G3P (with etymology pointing at Man, and chirality pointing to the Holy Spirit) has '$H_7$' in its acid form (Man's six waters to flow and the one likeness of this river ('*Euphrates*')). In its deprotonated form it has '$H_5$' representing the Holy Spirit working with all aspects that make Man fruitful (Christ-like). This is 'the four active water likenesses' and '*Euphrates*'. But because

'*Euphrates*' can also be applied to a long-term description then this can become centred on Man evangelising on other men. In this case the relevant water likenesses change as the internal pointing '*Gihon*' and '*Ethiopia*' no longer apply but '*Pison*', '*Havilah*', '*Hiddekel*', '*Assyria*' and '*Euphrates*' do – '$H_5$' still applies.

**O2** – The Holy Spirit uses two lights as well as the $PO_4$ and this is consistent with working with the Bible and Himself (in Hiddekel) and the greater and lesser lights (in Euphrates).

[xi] **1,3-bisphosphoglyceric acid ($C_3H_8O_{10}P_2$) ($C_3H_4O_2PO_4PO_4$) – 1,3-bisphosphoglycerate (($C_3H_4O_{10}P_2^{-4}$) is D-chiral:** Two light bringers bringing sweetness. A '2b' in the Calvin cycle.

**C3** – Conjugate acid: Christ's image life (1), Man's original spirit (2) and the regenerated aspect of Man's spirit (3). Conjugate base: 'body of Christ' (1); Man's regenerated spirit (2); and 'beast of the earth' (3).

**$PO_4$ (x2)** – The Holy Spirit is a light deliverer bringing lights from two light bringers using four vehicles – the regenerated Man, Jesus Christ, the Holy Spirit and the Bible (O4). The light bringers for these may be Jesus Christ or regenerated Man respectively. One bringing light into Man and the other Man giving light into the world or to other men.

**H8** – The conjugate: the original spirit of man contributes to 'Man's six waters to flow' – (H6). The working Holy Spirit by D-chirality uses the two likenesses of 'Gihon' to make it 'phosphoglyceric': 'Gihon' and the 'whole land of Ethiopia' (H2); H8 in total. The conjugate base: show all the Christ-centred changes to Man's original spirit; 'Havilah' and 'Ethiopia' and their part in 'Pison' and 'Gihon' – $H_4$.

**O2** – 'The two lights are consistent with the Holy Spirit working with the greater (light going in) and lesser (light going out) lights.

[xii] **Citric acid ($C_6H_8O_6$) and citrate ($C_6H_5O_6^{3-}$) is achiral** – the 'goodly tree' package of the regenerated man, 'good gold' and 'bdellium'. This is Christ centred and also centred on regenerated Man. A '1a' in Krebs cycle.

**C6** – The conjugate acid uses: 'God's six image lives given'. The conjugate base adds to or changes: 'Man's six image lives'.

**H8 to H5** – The conjugate acid represents what forms the 'goodly tree' and the goodly tree contains by etymology, 'the three resources' and is not just a 'one-likeness Pison' – but a 'three-likeness Pison' – in that the three resources are needed to 'live up to' its etymological name – $H_3$. As this includes 'Christ, the Holy Spirit and regenerated Man' it requires 'God's five waters to flow' – $H_8$ in total. Citrate reveals the likeness count: 'the five likenesses of Pison' and citrate is H5 in consistency with this.

**O7** – The Holy Spirit as a 'goodly tree' will contribute to all the light for pre-salvation, after salvation and long-term after salvation – 'Man and God's seven lights'.

[xiii] **Cis-aconitic acid ($C_6H_6O_6$) and cis-aconitate ($C_6H_3O_6^{3-}$) is achiral** and a '1b'– the poison from the darts – the Holy Spirit and Man together in their role having a sinful nature removed before and after salvation and the washing of his sinful desires.

**C6** – The conjugate acid uses: 'God's six image lives given'. The conjugate base adds to or changes: 'Man's six image lives'

**H6** – Conjugate acid: 'Man making six living water images' – H6. Conjugate base: Man's spirit being washed by 'the three resource likenesses of Pison' – H3.

**O6** – 'Man and God's six lights'. The negative charges are associated with oxygen atoms and this is consistent with the Holy Spirit working with lights.

[xiv] **Isocitric acid ($C_6H_8O_6$) and isocitrate ($C_6H_5O_6^{3-}$) is D chiral** an a '2a' – the goodly tree **now** from a Holy Spirit perspective (by chirality) and a whole spirit of Man perspective and what he contributes to the 'goodly tree' – in the events at Gihon

**C6** – The conjugate acid uses: 'God's six image lives given'. The conjugate base adds to or changes: 'Man's six image lives'.

**H8** – The conjugate acid again represents what forms the 'goodly tree' and the goodly tree contains by etymology, 'the three resources' and is not just a 'one-likeness Pison' – but a 'three-likeness Pison' – the three likenesses are needed to 'live up to' its etymological name – $H_3$. As this includes 'Christ, the Holy Spirit and regenerated Man' it requires 'God's five waters to f ow' – $H_8$ in total.

Isocitrate reveals the likeness count: 'the five likenesses of Gihon' and citrate is H5 in consistency with this ('Gihon', 'Ethiopia, 'good gold', 'bdellium' and 'onyx stone')

**O7** – The Holy Spirit as a 'goodly tree' will contribute to all the light for pre-salvation, after salvation and long-term after salvation – summarised as 'Man and God's seven lights'.

[xv] **Oxalosuccinic acid ($C_6H_6O_7$) – oxalosuccinate ($C_6H_3O_7^{3-}$) is L-chiral (Gihon)** – acid from tree amber – the original spirit pf Man being changed to be like Christ and carry Christ's image life and existing in the body of Christ – represented by the L-chirality.

**C6** – The conjugate acid uses: 'God's six image lives given'. The conjugate base adds to or changes: 'Man's six image lives'.

**H6 – H3:** The conjugate acid: 'Man making six living water images' – H6. The conjugate base will represent all likenesses associated with being Christ-like – 'good gold', 'Havilah' and 'Ethiopia'.

**O7** – 'Man and God's seven lights'. The three electrons will hover over oxygen atoms to represent the Holy Spirit working with lights – 'Man and God's seven lights'.

[xvi] **A-ketoglutaric acid ($C_5H_6O_5$) – α-ketoglutarate ($C_5H_4O_5^{-2}$) is achiral** – the alpha and the vinegar glued – Man's original spirit in a role as becoming a part of (or glued to) the body of Christ, and the formation of the 'cooperative'.

**C5** – This can be considered from before the addition of more of Christ's spirit (conjugate acid) or after addition of Christ's spirit (conjugate base). From the original spirit of Man's perspective: conjugate base – Man's five image lives (this contains Man's original spirit); conjugate acid – Man's six image lives (but no longer containing Man's original sinful spirit as it is now being added to or changed by Christ's image and light).

**H6 – H4:** The conjugate acid: 'Man making six living water images' – H6. The conjugate base will represent all likenesses associated with Man's original spirit (or 'whole land'): 'Pison', 'whole land of Havilah', 'Gihon' and 'whole land of Ethiopia'.

**O5** – The saved Man with regard to gifts of the spirit contributes to or requires 'the five standard after-salvation lights'. The two electrons will hover over oxygen atoms to represent the Holy Spirit working with lights.

Please note that, because of chirality, isocitrate can represent the Holy Spirit, and oxalosuccinate can represent the bride of Christ, and by this method these two can be a '3'.

[xvii] **Succinic acid ($C_4H_6O_4$) and succinate ($C_4H_4O_4^{-2}$) are achiral** – the amber of a tree – Man in his role carrying the gifts of the spirit and this is Holy Spirit centred and also original spirit of Man centred, a '3'.

**C4** – For the acid: Man's whole spirit (1), the Holy Spirit's image life (2), Man's beast of the earth (3) and Man's regenerated spirit (4). For the base Man's whole spirit can be swapped for the 'bride'.

**H6** – 'Man making six living water images'. The conjugate base will represent all likenesses associated with Man's original spirit (or 'whole land'): 'Pison', 'whole land of Havilah', 'Gihon' and 'whole land of Ethiopia'.

**O4** – The saved Man, with regard to gifts of the spirit, contributes to or requires the lesser light (1), the greater light (2), the Bible (3) and the Holy Spirit (4) – 'Man's four lights'.

[xviii] **Fumaric acid ($C_4H_4O_4$) and fumarate ($C_4H_2O_4^{-2}$) are achiral** – smoke – the blessing and presence of God. The regenerated spirit of Man in his role carrying a blessing from God. It seems to be Holy Spirit centred but working with the regenerated Man, as it is representing the fruitful Man at work, a '4'.

**C4** – 'The four interactive image lives of Pison'.

**H4 – H2:** 'the four river heads – H4'; the regenerated spirit of Man has two individual actions that make water 'likenesses' – the '*onyx stone*' and its part in 'Havilah' and 'Ethiopia'.

**O4** – Fumaric acid uses the standard four after-salvation lights – the lesser light, the greater light, the Bible and the Holy Spirit – to manifest his blessing.

www.ingramcontent.com/pod-product-compliance
Lightning Source LLC
LaVergne TN
LVHW012335130926
847902LV00028B/399
* 9 7 8 1 7 3 9 4 9 7 4 3 9 *